Statistical Physics of Dense Plasmas

Elementary Processes and Phase Transitions

Frontiers in Physics

Series Editor

Lou M. Chosen

The Eightfold Way
Murray Gell-Mann

Feynman Lectures On Gravitation
Richard Feynman, Fernando Morinigo, William Wagner, Brian Hatfield, David Pines

Feynman And Computation
Anthony Hey

The Physics Of Laser Plasma Interactions
William Kruer

Plate Tectonics
An Insider's History Of The Modern Theory Of The Earth
Naomi Oreskes

The Framework Of Plasma Physics
Richard D. Hazeltine, Francois L. Waelbroeck

Introduction To Ultrahigh Energy Cosmic Ray Physics
Pierre Sokolsky

Statistical Plasma Physics, Volume I
Basic Principles
Setsuo Ichimaru

Statistical Plasma Physics, Volume II
Condensed Plasmas
Setsuo Ichimaru

Computational Plasma Physics
With Applications To Fusion And Astrophysics
Toshi Tajima

The Field Theoretic Renormalization Group in Critical Behavior Theory and Stochastic Dynamics
A.N. Vasil'ev

Statistical Physics of Dense Plasmas
Elementary Processes and Phase Transitions
Setsuo Ichimaru

For more information about this series, please visit:
[www.crcpress.com/Frontiers-in-Physics/book-series/FRONTIERSPHYS]

Statistical Physics of Dense Plasmas

Elementary Processes and Phase Transitions

Setsuo Ichimaru

CRC Press
Taylor & Francis Group
Boca Raton London New York

CRC Press is an imprint of the
Taylor & Francis Group, an **informa** business

CRC Press
Taylor & Francis Group
6000 Broken Sound Parkway NW, Suite 300
Boca Raton, FL 33487-2742

Printed on acid-free paper

International Standard Book Number-13: 978-1-138-36466-0 (Paperback), 978-1-138-36468-4 (hardback)

Visit the Taylor & Francis Web site at
http://www.taylorandfrancis.com

and the CRC Press Web site at
http://www.crcpress.com

CONTENTS

PREFACE

In nature or in the laboratory setting, one encounters various physical systems containing numerous charged particles such as electrons and ions. These mobile charged particles interact with each other via long-range electromagnetic forces and thereby endow the system with medium-like properties. Such systems are called *plasmas.*

Physics of the plasma is concerned with the equilibrium and non-equilibrium properties of a statistical system containing many charged particles. One applies the basic principles of statistical mechanics and electrodynamics to elucidate the physical processes in such a system. The role that the long-range Coulomb interactions play in establishing collective phenomena, that is, organized behavior with strong inter-particle correlation, is particularly emphasized.

As the density is increased, the plasma begins exhibiting features characteristic of a condensed matter, where short-range as well as long-range forces conspire to bestow the plasma with the character of a *strongly coupled many-particle system.* As the temperature is lowered, quantum statistic and dynamic effects start to play dominant parts in plasmas, so that interplay with atomic, molecular, and nuclear physics becomes a significant issue. Associated with these is the emergence of features such as insulator-to-metal transition, order–disorder transition, paramagnetic to ferromagnetic transition, and chemical separation. Microscopic properties of the plasma depend delicately on these phase transitions, which in turn affect the macroscopic properties of the plasma through the rates of elementary and transport processes. Metallic hydrogen designates an issue most salient in these connections, experimentally as well as theoretically.

The present volume aims at elucidating basic issues in the statistical physics of dense plasmas, interfacing with condensed matter physics, atomic physics, nuclear physics, and astrophysics. Key phrases on the contents include: equations of state, phase transitions, thermodynamic properties, transport processes, radiative processes, and nuclear reactions, all influenced by the strong inter-particle, exchange and Coulomb correlation in dense plasmas.

Astrophysical dense plasmas are those we find in the interiors, surfaces, and outer envelopes of stellar objects such as neutron stars, white dwarfs, the Sun, and giant planets. Condensed plasmas in laboratory settings include those in ultrahigh-pressure metal physics experiments undertaken for the realization of metallic hydrogen. We recapitulate physics issues studied over the past several decades on the elementary processes and the phase transitions in such plasmas through the fundamental principles. Included in the recapitulation are: scattering of electromagnetic waves, injection of charged particles or X-ray, phase diagrams, resistivity, metallic hydrogen, stellar and planetary magnetism, and enhancement of thermonuclear as well as pycnonuclear reactions. Particularly crucial in these connections are the generation and emission of electromagnetic radiation and gravitational waves from the plasma surrounding neutron star systems, stellar black holes, and supermassive black holes; these imply the latest development of a foremost significance.

In completing this volume, the author expresses his heartfelt gratitude to R. Abe, J. Bardeen, G. Baym, D. M. Ceperley, L. B. Da Silva, R. Davidson, H. E. DeWitt, W. Ebeling, V. E. Fortov, J.-P. Hansen, T. Hatsuda, R. J. Hemley, W. B. Hubbard, N. Itoh, H. Iyetomi, B. Jancovici, H. Kitamura, W. Kohn, K. Makishima, J. Meyer-ter-Vehn, A. Nakano, S. Ogata, T. O'Neil, C. J. Pethick, D. Pines, R. Redmer, M. N. Rosenbluth, N. Rostoker, T. Tajima, S. Tanaka, H. Totsuji, K. Utsumi, H. M. Van Horn, P. Vashishta, J. Weisheit, and M. Yamada for collaboration and association over so many years. Parts of the volume depend to an extent on the lecture that the author delivered at an international conference in Brisbane, Australia, in December 2016.

Setsuo Ichimaru

Tokyo, Japan

1

INTRODUCTION

Plasmas are any statistical systems containing mobile charged particles. When such a system is condensed, interaction between particles becomes so effective that the system may undergo changes in the internal states or the phase transitions. One applies the basic principles of statistical mechanics to elucidate the thermodynamic properties and the rates of elementary processes in such a system. We begin this volume by surveying salient examples of dense plasmas in the astrophysical and laboratory settings.

1.1 DENSE PLASMAS IN NATURE

Astrophysical dense plasmas are those we find in the interiors, surfaces, and outer envelopes of stellar objects such as neutron stars, white dwarfs, the Sun, brown dwarfs, and giant planets (e.g., Van Horn, 1991; Ichimaru, 2004b). Condensed plasmas in the laboratory setting include: metals and alloys (solid, amorphous, liquid, and compressed), semiconductors (electrons, holes, and their droplets), various realizations of dense plasmas (shock-compressed, diamond-anvil cell, metal vaporization, pinch compression), and cryogenic, nonneutral plasmas (Davidson, 1990) including pure electron- or ion-plasmas (Driscoll & Malmberg, 1983; Bollinger et al., 1990) in the electromagnetic traps or on the surfaces of dielectrics such as liquid helium (Grimes, 1978).

The physics issues in such dense plasmas are (Ichimaru, Iyetomi, & Tanaka, 1987): phase transitions, construction of the phase diagrams, and accounting for the stellar as well as magnetic structures. Phase transitions to be considered are: gas to liquid, liquid to solid (Wigner, 1935, 1938), insulator to metal (Wigner & Huntington, 1935), hadrons to quark–gluon plasmas (Yagi, Hatsuda, & Miake, 2005), and para- to ferromagnetism (e.g., Landau & Lifshitz, 1960a).

Elementary processes involved in those plasmas then include (Ichimaru & Ogata, 1995): scattering of electromagnetic waves (Rosenbluth & Rostoker, 1962;

Ichimaru, 1973), photon transfers and opacities, emission of latent heat through phase transitions, electric and thermal transports, shear moduli of the crystalline solids, and enhanced thermonuclear as well as pycnonuclear reactions (Gamow & Teller, 1938; Cameron, 1959). The rates of these processes may depend sensitively on the changes in microscopic, macroscopic, thermodynamic, dielectric, and/or magnetic states of the matter. These changes of states may be associated with freezing transitions, chemical separations between the compositions, ionization or insulator-to-metal transitions, magnetic transitions, and transitions between normal to superconductive phases.

1.1.1 ASTROPHYSICAL DENSE PLASMAS

Interiors of the *main sequence stars* such as the *Sun* are dense plasmas constituted mostly of hydrogen. The Sun has a radius, $R_S \cong 6.69 \times 10^5$ km, and a mass, $M_S \cong 1.99 \times 10^{30}$ kg; the mass density is 1.41 g/cm^3 on average. The central part of the Sun has a mass density of approximately 1.56×10^2 g/cm^3, a temperature of approximately 1.5×10^7 K, and a pressure of approximately 3.4×10^5 Mbar (Bahcall & Pinsonneault, 1995). The mass fraction of hydrogen takes on a value of 0.36 near the center and 0.73 near the surface. The rates of nuclear reactions, photon transport and opacities, conductivities, atomic states, and their miscibility are all essential elements in setting a model for the Sun (Bahcall et al., 1982; Bahcall & Pinsonneault, 1995). The solar luminosity of $L_S \cong 3.85 \times 10^{26}$ W is to be accounted for, in particular, by the rates of nuclear reactions such as proton–proton chain reactions.

The interiors of giant planets (Jupiter, Saturn, Uranus, Neptune) offer important objects of study in the dense plasma physics (e.g., Hubbard, 1980, 1984; Stevenson, 1982). Typically, *Jupiter* has a radius, $R_J \cong 7.14 \times 10^4$ km $\cong 0.103\ R_S$, and a mass, $M_J \cong 1.90 \times 10^{27}$ kg $\cong 0.95 \times 10^{-3}\ M_S$. Figure 1.1 exhibits Jovian model showing three different internal phases: outer molecular hydrogen-helium fluid, inner metallic hydrogen-helium fluid, and the central fluid or solid "rock" composed of impurities. Models for the internal

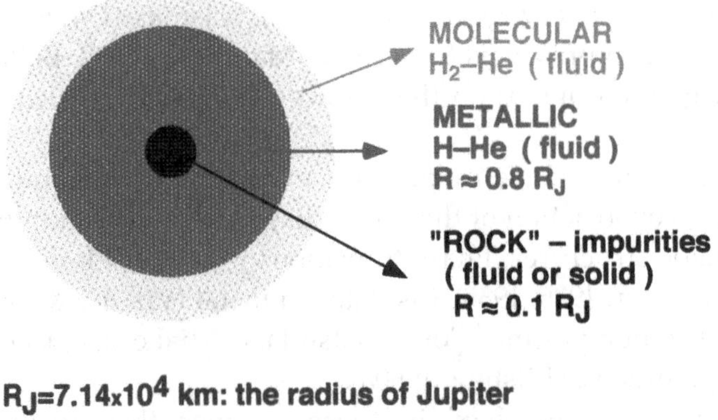

FIGURE 1.1 Jovian model showing three different phases.

structures of those planets were proposed on the bases of the thermodynamic and transport properties of the interiors, the surfaces, and the atmosphere coupled with the observational data such as gravitational harmonics (Hubbard & Marley, 1989); we particularly note precise measurements made by NASA's *Juno* spacecraft (Fortney, 2018).

The dominant magnetic-field contribution of the planet Jupiter for the external observer is the dipole of magnitude 4.2 gauss$\cdot R_J^3$ and a tilt of ~10° to the rotation axis (Smith, Davis Jr., & Jones, 1976). Closer to the planet, however, the multipole contributions are so large that an additional dipole term at a depth of ~2×10^4 km seems to be implied (Elphic & Russel, 1978). One must accurately assess the electric resistivity, in particular, for the internal metallic hydrogen-helium plasmas to account for the stellar magnetism.

The visible luminosity of the bright planet Jupiter, in fact, originates from solar radiation reflected from its surface, with albedo at 0.35. Jupiter has been known to emit radiation energy in the infrared range, approximately 2.7 times as intense as the total amount of radiation that it receives from the Sun. Through observation over terrestrial atmospheric transmission windows at 8–14 μm (Menzel, Coblentz, & Lampland, 1926) and 17.5–25 μm (Low, 1966), we have come to accept Jupiter being an unexpectedly bright infrared radiator. This feature has been reconfirmed quantitatively by a telescope airborne at an altitude of 15 km and through flyby measurements with *Pioneer 10* and *Pioneer 11* spacecrafts. The effective surface temperature determined from integrated infrared power over 8 to 300 μm was 129 ± 4 K, while the surface temperature calculated from equilibration with the absorbed solar radiation was 109.4 K (Hubbard, 1980). The balance needs to be accounted for by internal power generation; hence, the issue of *excess infrared luminosity* of Jupiter.

Returning to stellar objects, we now treat various stages of stellar evolution exhibited in Figure 1.2.

One of the proton–proton chains, the fundamental nuclear processes in the main-sequence stars, consists of the series of reactions,

$$p(p,e^+\nu_e)\,d\,(p,\gamma)^3\mathrm{He}\left({}^3\mathrm{He},2p\right){}^4\mathrm{He},$$

which altogether yields

$$4p \to \alpha + 2e^+ + 2\nu_e + 26.2\,(\mathrm{MeV}).$$

Here, p, d, α, e^+ and v_e, respectively, denote a proton, a deuterium, an alpha particle (the nucleus of ^{4}He), a positron, and an anti-neutrino; $p\ (p, e^+v_e)\ d$, for instance, describe the reactions,

$$p + p \to d + e^+ + \nu_e.$$

We parenthetically remark that the probabilities of these reactions are extremely small because of the involvement of charge transfers by positrons.

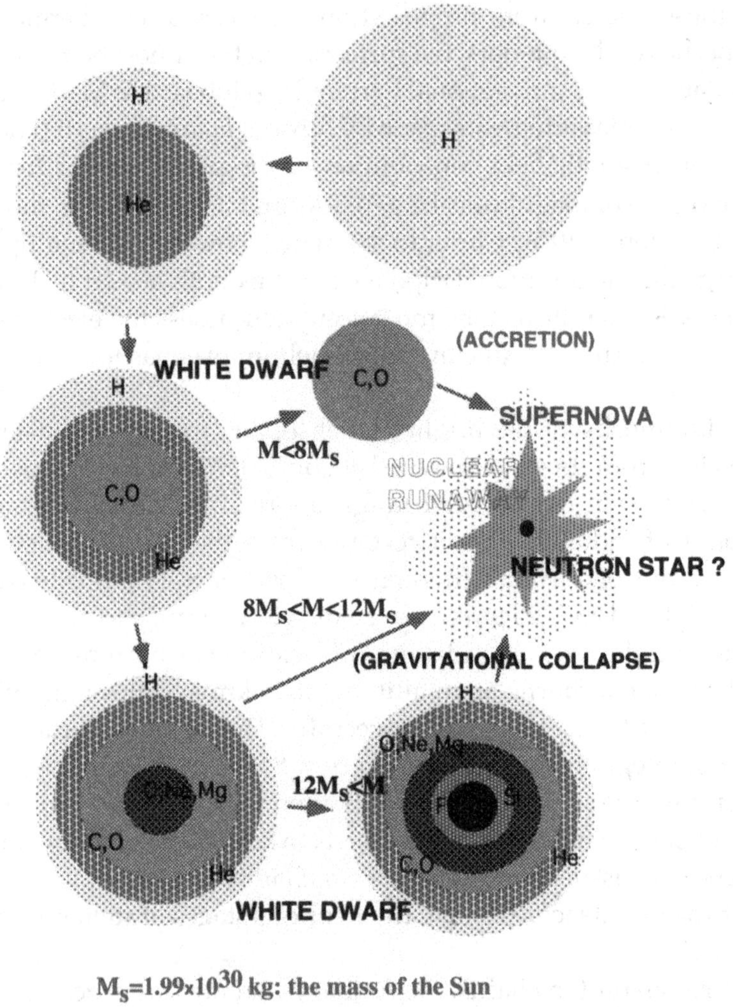

FIGURE 1.2 Schematics showing stages of stellar evolution.

Helium so produced is accumulated in the stellar core, and thereby leads the system to an inflated, relatively low-temperature star called a *red giant*. Helium burning, one of the major reaction processes in stellar evolution, then produces a carbon out of three α particles; another α capture in carbon then produces oxygen, and so on.

The *white dwarf* (Shapiro & Teukolsky, 1983) represents a final stage of stellar evolution, corresponding to a star of about one solar mass compressed to a characteristic radius of 5000 km and an average density of 10^6 g/cm^3. Its interior consists of a multi-ionic condensed matter composed of C and O as the main elements and Ne, Mg, …, Fe as trace elements. Condensed matter issues in white dwarfs include assessment of the possibilities of chemical separation or the phase diagrams associated with the freezing transitions of the multi-ionic plasmas (Stevenson, 1980; Van Horn, 1991; Ogata et al., 1993). These issues are related to the internal structures and cooling rates of white dwarfs (D'Antona & Mazzitrlli, 1990) as well as the rates of nuclear reactions (Ichimaru, 1993), evolution and nucleosynthesis (Clayton, 1968), detailed mechanisms of supernova explosion, and possible formation of a neutron star (Canal, Isern, & Labay, 1990; Nomoto & Kondo, 1991)

As a progenitor of type Ia supernova, a white dwarf with an interior consisting of a carbon–oxygen mixture can be considered a kind of binary-ionic mixture (BIM), with a central mass density of 10^7 to 10^{10} g/cm^3 and a temperature of 10^7 to 10^9 K (Starrfield et al., 1972; Whelan & Iben, 1973; Canal & Schatzman, 1976). Thermonuclear runaway leading to supernova explosion is expected to take place when the thermal output due to nuclear reactions exceeds the rate of energy losses. Assuming that neutrino losses are the major processes in the latter, one estimates (e.g., Arnett & Truran, 1969; Nomoto, 1982) that the nuclear runaway should take place when the nuclear power generated exceeds 10^{-9}~10^{-8} W/g. These values give approximate measures against which the rates of nuclear reactions may be compared.

The *neutron star* (e.g., Baym & Pethick, 1975; Shapiro & Teukolsky, 1983), another of the final stages in the stellar evolution, is a highly degenerate star corresponding approximately to a compression of a solar mass into a sphere with a radius of approximately 10 km.

We depict in Figure 1.3 a schematic structure of a neutron star. According to model calculations, it has an *outer crust*, consisting mostly of iron, with a thickness of several hundred meters and a mass density in the range of 10^4~10^7 g/cm^3. At these densities, iron atoms are completely ionized, so each contributes 26 conduction electrons to the system. At temperatures near 10^7 K, the thermal de Broglie wavelengths of the resultant Fe nuclei are substantially shorter than the average inter-nuclear separations; the iron nuclei may be regarded as forming classical ionic plasmas.

When the mass density exceeds a critical value near 10^7 g/cm^3 for the electron captures, neutron-rich "inflated" nuclei begin to emerge. At approximately 4×10^7 g/cm^3,

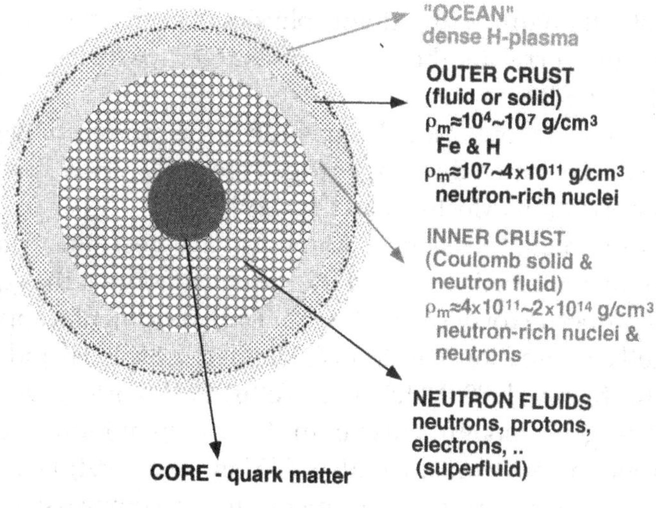

FIGURE 1.3 Structures of a neutron star.

the neutron drip density, the estimated atomic and mass numbers for such nuclei are $Z=36$ and $A=118$; at approximately 2×10^{14} g/cm^3, which defines the inner edge of an inner crust, one calculates $Z=201$ and $A=2500$ (Baym, Pethick, & Sutherland, 1971).

Over the bulk of the crustal parts, the nuclei are considered to form a *Coulomb solid*. A neutron star may then be looked upon as a "three-component star," consisting of an ultra-dense interior of neutron fluids with fractional constituents of proton and electrons, a crust of Coulomb solids, and a thin layer of "ocean" fluids. Electron transports and photon opacities in the outer crust and in the surface layer play the crucial parts (Gudmundsson, Pethick, & Epstein, 1982) in the estimate of the cooling rates for neutron stars (Nomoto & Tsuruta, 1981). Non-vanishing shear moduli associated with the crustal solids (Fuchs, 1936; Ogata and Ichimaru, 1990) lead to a prediction of rich spectra in the oscillations of a neutron star (McDemott, Van Horn, & Hansen, 1988; Strohmayer et al., 1991).

In the central core of a neutron star with a mass density in excess of 1 billion ton/cm^3, a phase with the *quark–gluon plasmas* is expected (Baym, 1995; Yagi, Hatsuda, & Miake, 2005).

1.1.2 DENSE PLASMAS IN LABORATORIES

The *inertial-confinement fusion* (ICF) is one of the schemes proposed for the development of nuclear fusion devices. It employs high-power laser beams to compress and implode a pellet that contains fusion fuel of hydrogen isotopes (e.g., Motz, 1979; Hora, 1991). The leading device currently in operation is National Ignition Facilities (NIF) at the Lawrence Livermore National Laboratory.

The states of plasmas for the ICF research resemble those of the solar interior mentioned in the preceding subsection; the projected temperatures are on the order of 10^7 to 10^8 K. Those materials that drive implosion of the fuel consist of high-Z elements, such as C, Al, Fe, Au, Pb, which after ionization form plasmas with charge numbers substantially greater than unity. The atomic physics of such a high-Z element is influenced strongly by the correlated behaviors of charged particles in a dense plasma (e.g., Goldstein et al., 1991).

Dense plasmas in the laboratories also include those produced by shock compression (e.g., Fortov, 1982), in pinch discharges (e.g., Pereira, Davis, & Rostoker, 1989), and through metal vaporization (Mostovych et al., 1991).

Ultrahigh-pressure metal physics experiments have been undertaken for laboratory realization of metallic hydrogen and for the elucidation of the equations of state and the transport properties of dense hydrogen. The experimental approaches include diamond-anvil-cell compression (e.g., Mao & Hemley, 1989, 1994) and shock compression (e.g., Dick and Kerley, 1980; Mitchell & Nellis, 1981; Fortov, 1995). Metallization of molecular hydrogen, though elusive in the diamond-anvil-cell experiments (Ruoff & Vanderbough, 1990; Mao, Hemley, & Hanfland, 1991; Hemley et al., 1996), was successfully demonstrated in experiments using compression through shock wave reverberation between electrically insulating sapphire (Al_2O_3) anvils (Weir, Mitchell, & Nellis, 1996; Da Silva et al., 1997).

Pressurized liquid metals offer an interesting environment in which to study nuclear reactions (Ichimaru, 1991, 1993). It may be shown that $d\,(p, \gamma)\,^3$He reactions can take place at a power-producing level on the order of a few kW/cm^3 if such a material is brought to a liquid-metallic state under an ultrahigh pressure on the order of 10^2 to 10^3 Mbar at a mass density of 10 to 10^2 g/cm^3 and a temperature of (1–2)$\times10^3$ K, in the vicinity of the estimated melting conditions for metallic hydrogen. Such a range of physical conditions may be accessible through extension of those ultrahigh-pressure metal technologies.

Some of the *nonneutral plasmas* (Davidson, 1990) cooled to sub-kelvin temperatures may likewise qualify as dense plasmas. Penning-trapped, pure electron (Driscoll & Malmberg, 1983), or pure ion (Bollinger et al., 1990) plasmas rotate around the magnetic axis due in part to space charge fields in the radial directions. In the frame co-rotating with the bulk of the particles, the Penning-trapped plasmas have been stably maintained at cryogenic temperatures, 10^{-2} to 10^0 K, for many hours, exhibiting ordered structures in the configurations of ions, reminiscent of a freezing transition.

1.2 BASIC PARAMETERS

We model a plasma at a temperature T as consisting of atomic nuclei (which will be called "ions") with an electric charge Ze and a rest mass M (= Am_N) and electrons with the electric charge $-e$ and the rest mass m. Here, Z is the charge number, and A refers to the mass number with m_N denoting the mass of a nucleon.

In certain cases, salient features of the plasma can be clarified through the study of a *one-component plasma* (OCP) (Ichimaru, 1982), as against a *two-component, electron–ion plasma* characterized in the foregoing paragraph. This model consists of a single species of charged particles with number density n embedded in a uniform background of neutralizing opposite charges.

Noticeable examples of such an OCP obeying the classical statistics may be found in the carbon ions in the interior of a white dwarf or in the iron ions in the outer crust of a neutron star, where dense electrons form the uniform negative-charge background.

Another example may be offered by the system of conduction electrons at metallic densities, where the ions forming crystalline lattice may be regarded as a smeared-out positive-charge background, in the so-called "jellium" model; the electrons are fermions with spin-1/2 obeying the quantum statistics.

Still another example may be offered by the system of protons in the metallic hydrogen, where the dense electrons may be regarded as a uniform negative-charge background; the protons, forming the OCP, are fermions with spin-1/2 obeying the quantum statistics.

1.2.1 CLASSICAL OCP

Consider the OCP obeying the classical statistics, as with the metallic ions in the white-dwarf interiors. The ratio between the average Coulomb energy, $(Ze)^2/a$, and the average kinetic energy, k_BT, per ion then introduces the *Coulomb coupling parameter* Γ for such an OCP via

$$\Gamma = \frac{(Ze)^2}{ak_BT} = 2.7\times10^{-5}Z^2\left[\frac{n}{10^{12}\ \text{cm}^{-3}}\right]^{1/3}\left[\frac{T}{10^6\ \text{K}}\right]^{-1}, \tag{1.1}$$

with $a=(4\pi n/3)^{-1/3}$ referring to the ion-sphere radius and k_B (= 1.38066×10^{-16} erg/K) denoting Boltzmann's constant; hereafter, lengths will be measured in units of a, unless specified otherwise. We call OCP as strongly coupled when $\Gamma>1$, weakly coupled when $\Gamma<0.1$, and intermediately coupled when $0.1\leq\Gamma\leq1$.

As we observe numerically in (1.1), Γ takes on extremely small values in ordinary gaseous plasmas. For example, we may assume $n=10^{11}$ cm^{-3}, $T=10^4$ K for a gaseous discharge plasma, $n=10^{16}$ cm^{-3}, $T=10^8$ K for a plasma in a controlled thermonuclear device, and $n=10^6$ cm^{-3}, $T=10^6$ K for a plasma in the solar corona. Assuming $Z=1$ for those plasmas, we find $\Gamma\approx10^{-3}$, 10^{-5}, and 10^{-7}, respectively. They are thus weakly coupled plasmas; their thermodynamic properties are analogous to those of an ideal gas.

Coulomb interaction plays the cardinal role in determining the physical properties of the plasma. In the theoretical treatment of plasmas in strong coupling, one cannot resort to a usual scheme of expansion in which the Coulomb interaction is regarded as a weak perturbation. We may also note that the interaction potential adopted for OCP has a simple and unique character: Among the repulsive potentials expressible as inverse power $r^{-\nu}$ of the distance r, OCP constitutes a typical example ($\upsilon=1$) of soft cases, while the hard-core case corresponds to the other extreme, $\upsilon\to\infty$.

It may therefore be said that we are here faced with a charged liquid problem. It is, nevertheless, to be noted that strongly coupled plasmas exhibit a remarkable similarity to hard-sphere systems in a number of significant aspects, such as short-ranged ordering in solidification (Alder & Wainwright, 1959). In fact, the short-ranged repulsive forces do play the essential parts as the origin of cohesive forces inducing Wigner crystallization (Wigner, 1935, 1938) and ferromagnetic transitions.

Thermodynamics and phases of the strongly coupled plasma, therefore, differ markedly from those of the weakly coupled plasma. We will find later an OCP may undergo the Wigner crystallization when Γ exceeds 172~180. We shall also find that the value of Γ critically affects the enhancement rate of nuclear reactions in dense plasmas (Ichimaru, 1993).

We may point out that the "dense plasmas" here refer to "high material-density plasmas," where $\Gamma>1$ mostly as n takes on an exceedingly large number. Under these circumstances, inter-particle correlations seriously affect the phase-related properties such as rates of the elementary processes (Ichimaru & Ogata, 1995).

In these connections, we also recognize the significance and importance of the activities in the fields of "high-energy density science." Here, the phase-related properties of high-energy density plasmas created as radiation-heated and shock-compressed matter are probed by powerful penetrating X-ray sources (Glenzer & Redmer, 2009; Dorchies & Recoules, 2016). It represents the warm matter or intense beam science, describing "high kinetic-energy density plasmas" mostly with $\Gamma < 1$. In this volume, we shall touch on some of those recent developments as well.

1.2.2 ELECTRON LIQUIDS AT METALLIC DENSITIES

In the jellium model of metals, the itinerant electrons are treated as forming a quantum liquid (Pines & Nozières, 1966), with the Fermi energy,

$$E_F = \frac{\hbar^2}{2m}\left(3\pi^2 n\right)^{2/3}, \tag{1.2}$$

measuring the average kinetic energy in such a system. Here, $\hbar = 1.0546 \times 10^{-27}$ erg·s denotes the Planck constant divided by 2π, $m = 9.1095 \times 10^{-28}$ g is the mass of an electron, and n stands for the number density of electrons. The electron being a fermion, each quantum state may be occupied by a single electron at most (Pauli exclusion principle); the quantum states are occupied by the n electrons successively from the lowest energy state. The Fermi energy corresponds to the energy of the final n-th electron.

Since E_F in (1.2) is an increasing function of n, $E_F \gg k_B T$ may be realized in an electron system at high density. In these circumstances, quantum effects as fermions become more dominant than thermal effects; it is thus relevant to use the Fermi energy as a measure of kinetic energies.

The ratio between the Coulomb energy and the Fermi energy is then given by

$$\frac{e^2 / a}{E_F} = 0.543 r_s, \tag{1.3}$$

with

$$r_s \equiv \frac{me^2 a}{\hbar^2}, \tag{1.4}$$

The thermodynamic properties of such an electron liquid are thus characterized by the two (density and temperature) dimensionless parameters (Ichimaru, 1982), that is, r_s in (1.4) and the degeneracy parameter,

$$\theta \equiv \frac{2mk_B T}{\hbar^2 \left(3\pi^2 n\right)^{2/3}}, \tag{1.5}$$

representing the ratio between the thermal energy, $k_B T$, and the Fermi energy, E_F. Here

$$a=\left(\frac{3}{4\pi n}\right)^{1/3} \tag{1.6}$$

is the *Wigner–Seitz radius* or the ion-sphere radius introduced earlier.

The r_s parameter, in fact, corresponds to this radius a measured in units of the Bohr radius, $a_B=\hbar^2/me^2=5.292\times10^{-9}$ cm. The values of r_s estimated for various metals at room temperatures are: 2.1 (Al), 2.7 (Mg), 3.3 (Li), 4.0 (Na), 5.0 (K), and 5.8 (Cs); in light of (1.3), they qualify as strongly coupled systems.

In these connections, it may also be useful to note the classical Coulomb coupling parameter (1.1) related to those dimensionless parameters as

$$\Gamma\equiv\frac{e^2}{ak_BT}=2\left(\frac{4}{9\pi}\right)^{2/3}\frac{r_s}{\theta}. \tag{1.7}$$

Thus, Γ, r_s, and θ are the dimensionless parameters of interest for dense plasmas.

1.3 CONSEQUENCES ON THE COULOMB INTERACTION

The extent of Coulomb coupling in plasmas depends on the temperature and the density. The Coulomb coupling parameter (1.1) gives a measure on the strength of such a coupling.

The consequences on the Coulomb coupling are multifold in plasmas. We here single out cooperative effects creating organized behaviors and collisional effects that may destroy the organized behaviors; we begin with the latter.

1.3.1 SCATTERING BY COULOMB FORCES

Consider an event of scattering between two charged particles, Z_1e and Z_2e, in the plasma, as depicted in Figure 1.4; their reduced mass and relative velocity are denoted as μ and v in the center-of-mass system. Following a standard treatment in the classical mechanics (e.g., Landau & Lifshitz, 1960b), one finds the relation between the scattering angle θ and the impact parameter b as

$$\cot\frac{\theta}{2}=\frac{b\mu v^2}{Z_1Z_2e^2}. \tag{1.8}$$

The differential cross-section dQ for scattering into an infinitesimal solid angle dΩ is then given by

$$\frac{dQ}{d\Omega}=\left[\frac{Z_1Z_2e^2}{2\mu v^2\sin^2(\theta/2)}\right]^2. \tag{1.9}$$

This is the formula for *Rutherford scattering.*

The increment of momentum in the direction of the OZ axis, in Figure 1.4, is $\Delta p = \mu v(1 - \cos\theta)$. Since the scattering is symmetric with respect to the OZ axis, the increments of momentum in the other two directions vanish on average. Consequently, the cross-section Q_m for the momentum transfer due to Coulomb scattering is calculated as

$$Q_m = \int_{\theta_{min}}^{\pi} (1 - \cos\theta)\left[\frac{Z_1 Z_2 e^2}{2\mu v^2 \sin^2(\theta/2)}\right]^2 2\pi \sin\theta d\theta$$
$$= 4\pi\left[\frac{Z_1 Z_2 e^2}{\mu v^2}\right]^2 \ln\left[\frac{1}{\sin(\theta_{min}/2)}\right]. \tag{1.10}$$

We have deliberately written the lower limit of the integral in (1.10) as θ_{min}; if we let θ_{min} approach zero, Q_m would diverge logarithmically.

According to (1.8), a small scattering angle θ corresponds to a large impact parameter b. The divergence in (1.10) thus originates from cumulative effects of large-distance scattering, reflecting the long-range nature of the Coulomb interaction.

When Z_1e and Z_2e are separated at a large distance in a plasma, many other plasma particles may be found between them; hence, the effective interaction between Z_1e and Z_2e is altered because of the presence of those other particles. In other words, in the calculation of the cross-section such as (1.10), one must take account of the effective potential including the many-body effects of other charged particles.

1.3.2 DEBYE SCREENING

Consider a weakly coupled ($\Gamma \ll 1$) classical OCP and treat the effective electrostatic potential, $\phi(\mathbf{r})$, around a test charge Z_0e introduced therein at the origin, $\mathbf{r} = 0$. The Poisson equation then is

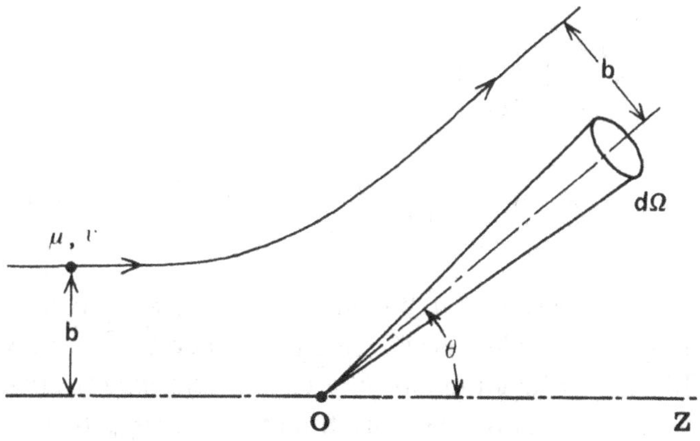

FIGURE 1.4 Coulomb scattering.

$$\nabla^2\phi(\mathbf{r}) = -4\pi Z_0 e\delta(\mathbf{r}) - 4\pi Ze\langle\rho(\mathbf{r})\rangle, \tag{1.11}$$

where $Ze\langle\rho(\mathbf{r})\rangle$ refers to the induced charge density and $\delta(\mathbf{r})$ is the three-dimensional delta-function representing the density distribution of a point particle located at the origin (cf. Appendix I).

The deviation $\langle\rho(\mathbf{r})\rangle$ from the average induced by the effective potential $\phi(\mathbf{r})$ is calculated in accord with the Boltzmann distribution as

$$\langle\rho(\mathbf{r})\rangle = n\exp\left[-\frac{Ze\phi(\mathbf{r})}{k_B T}\right] - n.$$

The argument of the exponential, being the ratio between the potential energy $Ze\langle\phi(\mathbf{r})\rangle$ and the thermal energy, takes on infinitesimal value when $\Gamma \ll 1$, so that we may expand it as

$$\langle\rho(\mathbf{r})\rangle \approx -\frac{Zen\phi(\mathbf{r})}{k_B T}.$$

Substitution of this expression in (1.11) yields a differential equation,

$$-\nabla^2\phi(\mathbf{r}) + \frac{4\pi n(Ze)^2}{k_B T}\phi(\mathbf{r}) = 4\pi Z_0 e\delta(\mathbf{r}), \tag{1.12}$$

which we may solve through the method of Fourier transformation, as explained in Appendix II, to obtain

$$\phi(\mathbf{r}) = \frac{Z_0 e}{r}\exp\left(-\frac{r}{\lambda_D}\right). \tag{1.13}$$

The parameter,

$$\begin{aligned}\lambda_D &\equiv \left[\frac{k_B T}{4\pi n(Ze)^2}\right]^{1/2} \\ &= 6.90\times 10^{-3} Z^{-1}\left[\frac{n}{10^{12}\ \mathrm{cm}^{-3}}\right]^{-1/2}\left[\frac{T}{10^6\ \mathrm{K}}\right]^{1/2} \ (\mathrm{cm}),\end{aligned} \tag{1.14}$$

introduced here is the *Debye length*, beyond which the electrostatic potential of the external charge may be regarded as effectively screened by the space charge induced in the plasma. This is the phenomenon called *Debye screening* (Debye & Hückel, 1923).

In light of these observations, we may thus take the θ_{min} ($\ll 1$, for a plasma with $\Gamma \ll 1$) in (1.10) as that determined from (1.8) in which b is set at λ_D, and find

$$Q_m = 4\pi\left[\frac{(Ze)^2}{\mu v^2}\right]^2 \ln\left(12\pi\lambda_D^3 n\right). \tag{1.15}$$

The logarithmic factor appearing here is called the *Coulomb logarithm.*

Argument of the Coulomb logarithm corresponds to an average number of particles in a sphere with a radius λ_D, that is, the *Debye number* $N_D = (4\pi/3)\lambda_D^3 n$, a huge number with $\Gamma \ll 1$. Numerically, that Coulomb logarithm takes on the value,

$$\ln(9N_D) = 16.3 + 1.15\log\left\{Z^{-6}\left[\frac{n}{10^{12}\ \text{cm}^{-3}}\right]^{-1}\left[\frac{T}{10^6\ \text{K}}\right]^3\right\}. \tag{1.16}$$

Hence we find that not only $N_D \gg 1$, but also $\ln(9N_D) \gg 1$ holds true.

The Debye screening thus represents an illuminating example of the *cooperative phenomena,* in that so many ($\sim N_D$) particles act together in the same (screening) directions by an infinitesimal degree each ($\sim N_D^{-1}$) to screen the effect of externally applied field.

1.3.3 THE ION-SPHERE MODEL

Thus far we have considered the problem of effective potential in a weakly coupled plasma and arrived at the concept of Debye screening. As we move into the strong coupling domain, $\Gamma > 1$, the Debye number becomes smaller than unity; the concept of Debye screening as a cooperative phenomenon is no longer applicable. A charged particle creates a sizable domain around itself where no other particles are likely to be found, a sort of a territorial domain of its own influence, which may be looked upon as a *Coulomb hole.*

To understand salient features of such a strongly coupled plasma, it is instructive to introduce the *ion-sphere model* (Salpeter, 1954), which is equivalent to the Wigner–Seitz sphere used in the solid-state physics (e.g., Pines, 1963). As Figure 1.5 illustrates, one considers a charged particle Ze and a surrounding neutralizing charge sphere, whose total electric charge is just to cancel the charge Ze. This sphere thus represents the terrestrial domain of influence for the charge Ze. Its radius is a of (1.6); the charge density, $-3Ze/4\pi a^3$.

The ion sphere thus consists of a single ion and its surrounding negative-charge sphere. The electrostatic energy E_{IS} associated with the ion sphere is then calculated in the following way: First, the electrostatic potential produced by the negative-charge sphere at r ($\leq a$) is

$$\phi_{IS}(r) = -\frac{3Ze}{4\pi a^3}\int_{r'\leq a} d\mathbf{r}'\frac{1}{|\mathbf{r}-\mathbf{r}'|} = -\frac{3Ze}{2a} + \frac{Ze}{2a}\left(\frac{r}{a}\right)^2. \tag{1.17}$$

The electrostatic energy of the negative-charge sphere itself is then

$$-\frac{1}{2}\int_{r'\leq a} d\mathbf{r}'\frac{3Za}{4\pi a^3}\phi_{\text{IS}}(r)=\frac{3}{5}\frac{(Ze)^2}{a}.$$

Summation of this energy and $Ze\phi_{\text{IS}}(r)$ yields

$$\frac{E_{\text{IS}}}{k_BT}=-0.9\Gamma+0.5\Gamma\left(\frac{r}{a}\right)^2. \tag{1.18}$$

The first term on the right-hand side of (1.18) represents the electrostatic energy when the ion Ze is located at the center of the ion sphere. The second term, being proportional to r^2, induces a motion of the harmonic-oscillator type to the ion. The average energy of a harmonic oscillator, including the kinetic energy, is k_BT per a degree of freedom. The density of internal energy U_{IS}, calculated on the basis of the ion-sphere model, is thus

$$\frac{U_{\text{IS}}}{nk_BT}=-0.9\Gamma+3. \tag{1.19}$$

This result, in fact, takes on a value close to computer simulation results for the strongly coupled OCP, as we shall see later.

According to the ion-sphere model of Figure 1.5, the electrostatic potential of the central charge is confined within a distance $\alpha_1 a$, where α_1 is a correction factor of order unity accounting for uncertainty involved in a strict enforcement of the ion-sphere model. Putting $Z_1=Z_2=Z$ in (1.8), we determine the scattering angle $\theta_{\min}$ corresponding to the impact parameter at $\alpha_1 a$. The cross-section (1.10) so determined takes the form,

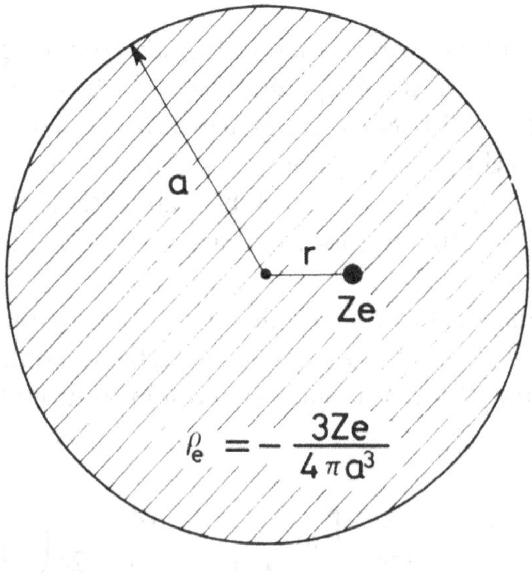

FIGURE 1.5 The ion-sphere model.

$$Q_m = 2\pi\left(\frac{Z^2e^2}{\mu v^2}\right)^2 \ln\left\{1+\left(\frac{\alpha_1 a\mu v^2}{Z^2e^2}\right)^2\right\}. \tag{1.20}$$

When $\Gamma > 1$, we may take $\alpha_1 a\mu v^2/(Ze)^2 < 1$ for the bulk of the particles, so that the logarithm in (1.20) may be expanded to yield

$$Q_m \simeq 2\pi(\alpha_1 a)^2. \tag{1.21}$$

As we would have anticipated, the scattering cross-section (1.21) is here seen to be proportional to the cross-section πa^2 of the ion sphere.

1.3.4 PLASMA OSCILLATION

Thus far we have considered the issues of static variation in plasmas such as Debye screening and the ion-sphere model. The Coulomb interaction produces not only such a static cooperative effect, but also the *plasma oscillation*, which may be regarded as dynamic cooperative phenomenon.

To treat such a problem, we begin with the equation of motion describing the temporal behavior of density fluctuation in an OCP, consisting of N particles with electric charge Ze and mass m in volume V; $n = N/V$ is then the average number density. Let $\mathbf{r}_j(t)$ be the spatial trajectory of the j-th particle; the density fluctuation arising from this particle is then expressed as $\delta[\mathbf{r} - \mathbf{r}_j(t)]$, in terms of the delta-function introduced in (1.11). We may thus express spatial fluctuations of the density field as

$$\rho(\mathbf{r},t) = \sum_{j=1}^{N} \delta\left[\mathbf{r} - \mathbf{r}_j(t)\right], \tag{1.22}$$

summing up the contributions from all the particles. We now calculate the Fourier component of the density fluctuation with wave vector $\mathbf{k}$, with the periodic boundary conditions appropriate to volume V (cf. Appendix II), to yield

$$\begin{aligned} \rho_{\mathbf{k}}(t) &= \int d\mathbf{r}\left\{\sum_{j=1}^{N} \delta\left[\mathbf{r} - \mathbf{r}_j(t)\right] - n\right\}\exp(-i\mathbf{k}\cdot\mathbf{r}) \\ &= \sum_{j=1}^{N} \exp\left[-i\mathbf{k}\cdot\mathbf{r}_j(t)\right] - N\delta(\mathbf{k},0). \end{aligned} \tag{1.23}$$

This quantity refers to the spatial Fourier components of density fluctuations; $\delta(\mathbf{k}, \mathbf{k}')$ is the three-dimensional Kronecker delta.

We then derive an equation of motion by differentiating this quantity twice with respect to time; since $\dot{\mathbf{r}}_j = \mathbf{v}_j$ represents the velocity of the j-th particle, we find

$$\ddot{\rho}_{\mathbf{k}} = -\sum_j \left\{ (\mathbf{k}\cdot\mathbf{v}_j)^2 + i\mathbf{k}\cdot\dot{\mathbf{v}}_j \right\} \exp(-i\mathbf{k}\cdot\mathbf{r}_j). \tag{1.24}$$

The quantity $\dot{\mathbf{v}}_j$ representing the acceleration of the j-th particle can be calculated from the force exerted upon this particle by all other charges and the neutralizing background charge,

$$\dot{\mathbf{v}}_j = -\frac{1}{m}\frac{\partial}{\partial \mathbf{r}_j}\sum_{l(\neq j)} \frac{(Ze)^2}{|\mathbf{r}_j - \mathbf{r}_l|} + \left(\text{acceleration by the background charge}\right).$$

Since the Coulomb potential is Fourier expanded as (cf. Appendix II),

$$\frac{Ze}{r} = \sum_{\mathbf{k}} \frac{4\pi Ze}{k^2} \exp(i\mathbf{k}\cdot\mathbf{r}), \tag{1.25}$$

we obtain from (1.24)

$$\ddot{\rho}_{\mathbf{k}} = -\sum_j (\mathbf{k}\cdot\mathbf{v}_j)^2 \exp(-i\mathbf{k}\cdot\mathbf{r}_j) - \frac{4\pi(Ze)^2}{m} {\sum_{\mathbf{q}}}' \frac{\mathbf{k}\cdot\mathbf{q}}{q^2} \rho_{\mathbf{k}-\mathbf{q}}\rho_{\mathbf{q}},$$

where $\sum'$ implies that the $\mathbf{q} = 0$ term is not to be summed in the summation. The first term on the right-hand side represents the effect of the translational motion of the individual particles; the second term stems from the Coulomb interaction. From the latter, we particularly single out the $\mathbf{q} = \mathbf{k}$ term to yield,

$$\ddot{\rho}_{\mathbf{k}} + \omega_p^2 \rho_{\mathbf{k}} = -\sum_j (\mathbf{k}\cdot\mathbf{v}_j)^2 \exp(-i\mathbf{k}\cdot\mathbf{r}_j) - \frac{4\pi(Ze)^2}{m} {\sum_{\mathbf{q}(\neq\mathbf{k})}}' \frac{\mathbf{k}\cdot\mathbf{q}}{q^2} \rho_{\mathbf{k}-\mathbf{q}}\rho_{\mathbf{q}}. \tag{1.26}$$

Here, we have introduced another important parameter,

$$\omega_p \equiv \left[\frac{4\pi n(Ze)^2}{m}\right]^{1/2}, \tag{1.27}$$

called the *plasma frequency*; for the electron plasma, it takes on the values,

$$\omega_p = 5.64 \times 10^{10} \left[\frac{n}{10^{12}\ \text{cm}^{-3}}\right]^{1/2} \ (\text{s}^{-1}). \tag{1.28}$$

If the right-hand side of (1.26) can be neglected altogether, the temporal variation of the density with wave vector $\mathbf{k}$ obeys an equation of motion for the harmonic oscillator with the frequency (1.27). As we see in (1.23), $\rho_{\mathbf{k}}$ contains coordinates of all the particles; nevertheless, we find here the possibility of such a quantity exhibiting an orderly oscillatory behavior under the action of the Coulomb forces, that is, an appearance of the *collective motion*.

1.3.5 COLLECTIVE MOTION AND INDIVIDUAL-PARTICLES BEHAVIOR

In the early 1950s, Bohm and Pines advanced a series of papers (Bohm & Pines, 1951, 1953; Pines & Bohm, 1952) dealing with "a collective description of electron interactions," in which they explicitly stated:

> The density fluctuations may be split into two approximately independent components. The collective component, that is, the *plasma oscillation*, is present only for wavelengths greater than the Debye length. The individual particles component represents a collection of *individual electrons surrounded by co-moving clouds of screening charges*; collisions between them may be negligible under the *random-phase approximation.*

The equation of motion (1.26) derived in the preceding section, in fact, corroborates precisely with this statement. To see this, let us investigate the relative magnitudes of the terms on its right-hand side in some detail.

First, we carry out a statistical estimate of the first term on the right-hand side of (1.26) with the aid of the Maxwell–Boltzmann velocity distribution,

$$f_B(\mathbf{v}) = \left(\frac{m}{2\pi k_B T}\right)^{3/2} \exp\left(-\frac{mv^2}{2k_B T}\right), \tag{1.29}$$

to obtain,

$$\langle \text{the first term} \rangle \approx (k_B T / m) k^2 \rho_{\mathbf{k}}. \tag{1.30}$$

The second term, on the other hand, represents a nonlinear term consisting of the products of density variations. On the complex-number plane as shown in Figure 1.6, the Fourier components of density fluctuations are expressed as summations of unit vectors over the total number of the particles. For a uniform system, the phase angles, $-\mathbf{k}\cdot\mathbf{r}_j$, distribute randomly as long as $\mathbf{k} \neq 0$; hence, the expectation value $\langle \rho_{\mathbf{k}} \rangle = 0$.

Bohm and Pines then assumed a possibility that an analogous situation may become applicable to the second term on the right-hand side consisting of products of density fluctuations. They thereby introduced an approximation, called the *random-phase approximation* (RPA), in which those product terms be ignored. Even though the phase $-\mathbf{k}\cdot\mathbf{r}_j$ may be distributed randomly, however, products between fluctuations, $\rho_{\mathbf{k}-\mathbf{q}}\rho_{\mathbf{q}}$, cannot generally be ignored. Setting the origin of the nomenclature aside, we may reinterpret the RPA as an approximation applicable to the cases of weak density variations whereby the equation of motion may be linearized with respect to fluctuations. The RPA thus provides a fairly accurate description of weakly coupled plasmas near thermodynamic equilibrium. In these circumstances, we may ignore the second term on the right-hand side of (1.26), as it consists of nonlinear terms of the fluctuations.

Let us thus adopt this RPA. Equation (1.26) now becomes

$$\ddot{\rho}_{\mathbf{k}} + \omega_p^2 \rho_{\mathbf{k}} = -(k_B T / m) k^2 \rho_{\mathbf{k}}. \tag{1.31}$$

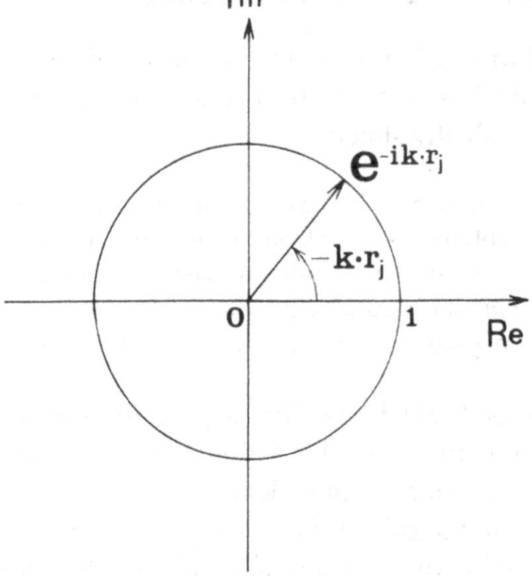

FIGURE 1.6 Representation of a particle on the complex plane.

In the long-wavelength regime such that $k^2 \ll k_D^2$, where

$$k_D = \left[\frac{4\pi n(Ze)^2}{k_B T}\right]^{1/2}, \tag{1.32}$$

called the *Debye wavenumber*, is the inverse of the Debye length (1.14), we may ignore the right-hand side of (1.31); the density fluctuations in plasmas behave collectively and oscillate at a frequency near ω_p.

In the short wavelength regime such that $k^2 \gg k_D^2$, Eq. (1.31) now becomes,

$$\ddot{\rho}_{\mathbf{k}} = -(k_B T / m)k^2 \rho_{\mathbf{k}}.$$

This is nothing but the equation of motion for a collection of individual particles screened within the Debye length by co-moving clouds of screening charges.

The statements of Bohm and Pines concerning the collective behaviors in plasmas thus hold true within the RPA.

2

FUNDAMENTALS

Charged particles in plasmas interact with each other via Coulomb forces and thereby exhibit a variety of interesting features as described in the preceding chapter. In many cases, these features are expressed in terms of density–density correlations and fluctuations. For example, the Debye screening and the ion-sphere model are related to static correlations; the plasma oscillation may be looked upon as an appearance of dynamic correlations. In this chapter, as the fundamentals on the statistical physics of dense plasmas, we study fluctuations, density–density correlations and their relationship with thermodynamic variables along with dielectric formulations, density-functional approaches, and computer simulation methods.

As a specific example of multicomponent plasmas, we treat here mostly the cases of an electron liquid such as the system of conduction electrons in the metal, where dependence on spin orientations as fermions is explicitly accounted for. Extension of those theories to other cases of multicomponent plasmas is rather straightforward.

2.1 DENSITY-FLUCTUATION EXCITATIONS

For simplicity and clarity, we begin with the second-quantization formalisms for a statistical system of identical particles with total number N in volume V, where the distinction between bosons and fermions is particularly singled out.

2.1.1 SYSTEM OF IDENTICAL PARTICLES

The properties of a quantum system consisting of many identical particles are most conveniently described in terms of the second-quantized, Heisenberg representation, particle-creation and annihilation operators (e.g., Kadanoff & Baym, 1962). The creation operator $\psi^{\dagger}(\mathbf{r},t)$, when acting to the right on the state of the system, adds a particle to the state at the space-time point $\mathbf{r}$, t; the annihilation operator $\psi(\mathbf{r},t)$, the adjoint of the creation operator, acting to the right, removes a particle from the state at the point $\mathbf{r}$, t.

The macroscopic operators of direct physical interest can all be expressed in terms of products of a few ψs and ψ†s. For example, the density of particles at the point **r**, t is

$$n(\mathbf{r},t) = \psi^{\dagger}(\mathbf{r},t)\psi(\mathbf{r},t), \tag{2.1a}$$

since the act of removing and then immediately replacing a particle at **r**, t measures the density of particles at that point; the operator for the total number of particles is

$$N(t) = \int d\mathbf{r}\psi^{\dagger}(\mathbf{r},t)\psi(\mathbf{r},t). \tag{2.1b}$$

Similarly, the total energy of a system of particles of mass m interacting through an instantaneous two-body potential v(r) is given by

$$\begin{aligned} H(t) = \hbar^2 \int d\mathbf{r} \frac{\nabla\psi^{\dagger}(\mathbf{r},t)\cdot\nabla\psi(\mathbf{r},t)}{2m} \\ + \frac{1}{2}\int d\mathbf{r}\int d\mathbf{r}'\psi^{\dagger}(\mathbf{r},t)\psi^{\dagger}(\mathbf{r}',t)v(|\mathbf{r}-\mathbf{r}'|)\psi(\mathbf{r}',t)\psi(\mathbf{r},t). \end{aligned} \tag{2.2}$$

The equation of motion for any operator in the Heisenberg representation is

$$\frac{i\partial X(t)}{\partial t} = [X(t), H(t)]. \tag{2.3}$$

Here and hereafter, [A, B] = AB − BA means a commutator. Since [H(t), H(t)] = 0, we see that the Hamiltonian is independent of time. Also, the Hamiltonian does not change the number of particles, [H, N(t)] = 0, and therefore N(t) is also independent of time. Because of the time independence of H, (2.3) may be integrated in the form

$$X(t) = \exp(iHt)X(0)\exp(-iHt). \tag{2.4}$$

Particles may be classified into one of two types: Fermi–Dirac particles, also called fermions, which obey the exclusion principle, and Bose–Einstein particles, or bosons, which do not. The wave function of any state of a collection of bosons must be a symmetric function of the coordinates of the particles, whereas, for fermions, the wave functions must be antisymmetric. One of the main advantages of the second-quantization formalism is that these symmetric requirements are simply represented in the equal-time commutation relations of the creation and annihilation operators. These commutation relations are

$$\begin{aligned} \psi(\mathbf{r},t)\psi(\mathbf{r}',t) \mp \psi(\mathbf{r}',t)\psi(\mathbf{r},t) = 0 \\ \psi^{\dagger}(\mathbf{r},t)\psi^{\dagger}(\mathbf{r}',t) \mp \psi^{\dagger}(\mathbf{r}',t)\psi^{\dagger}(\mathbf{r},t) = 0 \\ \psi(\mathbf{r},t)\psi^{\dagger}(\mathbf{r}',t) \mp \psi^{\dagger}(\mathbf{r}',t)\psi(\mathbf{r},t) = \delta(\mathbf{r}-\mathbf{r}') \end{aligned} \tag{2.5}$$

where the upper sign refers to Bose–Einstein particles and the lower sign refers to Fermi–Dirac particles. We see, for fermions, that $\psi^2(\mathbf{r}, t) = 0$. This is an expression of the exclusion principle—it is impossible to find two identical fermions at the same point in space and time.

2.1.2 STRUCTURE FACTORS AND CORRELATION ENERGY

Essential features of the density-fluctuation excitations may be described in terms of the structure factors (e.g., Ichimaru, 2004b), which represent the spectral distributions of fluctuations in space and time. The dynamic structure factor,

$$S(\mathbf{k},\omega) = \frac{1}{2\pi V}\int_{-\infty}^{\infty} dt \langle \rho_{\mathbf{k}}(t'+t)\rho_{-\mathbf{k}}(t')\rangle \exp(i\omega t) \tag{2.6}$$

defined in terms of the spatial Fourier transformation of density fluctuations,

$$\rho_k(t) = \int d\mathbf{r}\left[n(\mathbf{r},t) - N/V\right]\exp(-i\mathbf{k}\cdot\mathbf{r}), \tag{2.7}$$

portray the spectral distribution of the fluctuations in the wave vector and frequency space, $(\mathbf{k}, \omega)$. In (2.6) and hereafter, $\langle\cdots\rangle$ denotes average over states of the system.

The static structure factor $S(\mathbf{k})$ defined as

$$S(\mathbf{k}) = \frac{1}{N}\left\langle \left|\rho_{\mathbf{k}}(t)\right|^2 \right\rangle = \frac{1}{n}\int_{-\infty}^{\infty} d\omega S(\mathbf{k},\omega) \tag{2.8}$$

corresponds to the spectral distribution of spatial density fluctuations, which describes spatial density configurations such as lattice structures; $n = N/V$ is the number density on average.

The radial distribution function $g(\mathbf{r})$ is a joint probability density of finding two particles at a separation r. It is related directly to the static structure factor (2.8) as

$$g(\mathbf{r}) = 1 + \frac{1}{n}\int \frac{d\mathbf{k}}{(2\pi)^3}\left[S(\mathbf{k}) - 1\right]\exp(i\mathbf{k}\cdot\mathbf{r}). \tag{2.9}$$

The correlation energy U_{int} per unit volume, a statistical average of the second term on the right-hand side of (2.2), can then be calculated, once either $S(\mathbf{k})$ or $g(\mathbf{r})$ is known, through formulae,

$$\begin{aligned} U_{\text{int}} &= \frac{n}{2}\int \frac{d\mathbf{k}}{(2\pi)^3} v(k)\left[S(\mathbf{k}) - 1\right] \\ &= \frac{n^2}{2}\int d\mathbf{r} v(\mathbf{r})\left[g(\mathbf{r}) - 1\right]. \end{aligned} \tag{2.10}$$

Here, $v(k)$ is the Fourier transform of $v(\mathbf{r}) = e^2/r$; explicitly, it is $4\pi e^2/k^2$, as reproduced later with (2.19). Its derivation may be accounted for, in reference to (AII.13).

2.1.3 SYSTEM OF ELECTRONS AT METALLIC DENSITIES

We now turn to an electron liquid at a metallic density and consider the creation and annihilation operators, $c^{\dagger}_{\mathbf{p}\sigma}$ and $c_{\mathbf{p}\sigma}$, for the electrons with momentum $\mathbf{p}$ and spin σ (= ↑, ↓). The operators satisfy the anticommunicator relations for the fermions:

$$\left\{c^{\dagger}_{\mathbf{p}\sigma}, c^{\dagger}_{\mathbf{p}'\sigma'}\right\} = \left\{c_{\mathbf{p}\sigma}, c_{\mathbf{p}'\sigma'}\right\} = 0, \tag{2.11a}$$

$$\left\{c^{\dagger}_{\mathbf{p}\sigma}, c_{\mathbf{p}'\sigma'}\right\} = \delta_{\mathbf{p}\mathbf{p}'}\, \delta_{\sigma\sigma'} \tag{2.11b}$$

Here and hereafter, $\{A, B\} = AB + BA$, meaning an anticommutator; $\delta_{\mathbf{p}\mathbf{p}'}$ and $\delta_{\sigma\sigma'}$ are the Kronecker deltas.

For a treatment of the density-fluctuation excitations, it is convenient to work with operators representing electron-hole pairs (cf. Wigner, 1932; Brittin and Chappell, 1962),

$$\rho_{\mathbf{p}\mathbf{k}\sigma} = c^{\dagger}_{\mathbf{p}\sigma} c_{\mathbf{p}+\hbar\mathbf{k}\sigma}. \tag{2.12}$$

The Fourier component of spin-dependent, density-fluctuation excitations with wave vector $\mathbf{k}$ is then given by

$$\rho_{\mathbf{k}\sigma} = \sum_{\mathbf{p}} \rho_{\mathbf{p}\mathbf{k}\sigma}; \tag{2.13a}$$

total density fluctuations are calculated as

$$\rho_{\mathbf{k}} = \sum_{\mathbf{p},\sigma} \rho_{\mathbf{p}\mathbf{k}\sigma}. \tag{2.13b}$$

Its classical counterpart has been introduced in (1.23).

The Fourier components of the charge- and spin-density-fluctuation excitations are

$$\rho^{(c)}_{\mathbf{k}} = -e \sum_{\mathbf{p},\sigma} \rho_{\mathbf{p}\mathbf{k}\sigma}, \tag{2.14a}$$

$$\rho^{(s)}_{\mathbf{k}} = \sum_{\mathbf{p}} \left(\rho_{\mathbf{p}\mathbf{k}\uparrow} - \rho_{\mathbf{p}\mathbf{k}\downarrow}\right). \tag{2.14b}$$

The occupation number operator for the state $(\mathbf{p}, \sigma)$ is finally expressed as

$$n_{\mathbf{p}\sigma} = \rho_{\mathbf{p},\mathbf{k}=0,\sigma}, \tag{2.15}$$

as the act of removing and then immediately replacing a particle at **p** and σ measures the occupation number at that point.

2.2 DIELECTRIC FORMULATION

The dielectric response functions are a class of linear response functions that stem from the exchange and Coulomb interactions between particles. These describe the dielectric properties associated with various density-fluctuation excitations in plasmas (e.g., Ichimaru, 2004b).

Response functions of plasmas are formulated through the application of an external potential field $\Phi_\sigma^{\text{ext}}(\mathbf{k},\omega)$ that couples with the density fluctuations (2.12). The total Hamiltonian of the system is given as a sum of the unperturbed and external Hamiltonians,

$$H_{\text{tot}} = H + H_{\text{ext}}, \tag{2.16}$$

so that

$$H = \sum_{\mathbf{p}\sigma} \frac{p^2}{2m} c_{\mathbf{p}\sigma}^{+} c_{\mathbf{p}\sigma} + \frac{1}{2} \sum_{\substack{\mathbf{p},\mathbf{p}' \\ \mathbf{k}(\neq 0) \\ \sigma,\sigma'}} Z_\sigma Z_{\sigma'} v(k) c_{\mathbf{p}+\hbar\mathbf{k},\sigma}^{+} c_{\mathbf{p}'-\hbar\mathbf{k},\sigma'}^{+} c_{\mathbf{p}'\sigma'} c_{\mathbf{p}\sigma}, \tag{2.17}$$

$$H_{\text{ext}} = \sum_{\mathbf{p}\sigma} Z_\sigma e \Phi_\sigma^{\text{ext}}(\mathbf{k},\omega) c_{\mathbf{p},\sigma}^{+} c_{\mathbf{p}-\hbar\mathbf{k},\sigma} \exp(-i\omega t + 0t) + \text{hc}. \tag{2.18}$$

Here

$$v(k) = \frac{4\pi e^2}{k^2} \tag{2.19}$$

is the Fourier transform of the Coulomb interaction e^2/r; the summation in the second term of (2.17) implies omission of the terms with $\mathbf{k}=0$; and "hc" stands for the Hermitian conjugate. The "0" in (2.18) denotes a positive infinitesimal, which designates an adiabatic turning-on of the external potential; this precaution then ensures a *causal response* of the system.

We here consider a multicomponent system of fermions with electric charge Z_σ for generality; in the case of an electron system, one sets $Z_\sigma = -1$.

According to (2.3), the Heisenberg equation of motion for $\rho_{\mathbf{pk}\sigma}$ is

$$i\hbar \frac{\partial}{\partial t} \rho_{\mathbf{pk}\sigma} = \left[\rho_{\mathbf{pk}\sigma}, H_{\text{tot}}\right]. \tag{2.20}$$

Explicit calculation with the aid of (2.17) and (2.18) yields

$$i\hbar\frac{\partial}{\partial t}\rho_{\mathbf{pk}\sigma} = \hbar\omega_{\mathbf{pk}}\rho_{\mathbf{pk}\sigma} \tag{2.21a}$$

$$+\frac{1}{2}v(k)\left\{\rho_{\mathbf{k}}, n_{\mathbf{p}\sigma} - n_{\mathbf{p}+\hbar\mathbf{k},\sigma}\right\} \tag{2.21b}$$

$$+\frac{1}{2}\sum_{\mathbf{q}(\neq 0,\mathbf{k})} v(q)\left\{\rho_{\mathbf{q}}, \rho_{\mathbf{p},\mathbf{k}-\mathbf{q},\sigma} - \rho_{\mathbf{p}+\hbar\mathbf{q},\mathbf{k}-\mathbf{q},\sigma}\right\} \tag{2.21c}$$

$$+Z_\sigma\Phi_\sigma^{\text{ext}}(\mathbf{k},\omega)\left(n_{\mathbf{p}\sigma} - n_{\mathbf{p}+\hbar\mathbf{k},\sigma}\right)\exp(-i\omega t + 0t), \tag{2.21d}$$

where

$$\omega_{\mathbf{pk}} = \frac{\mathbf{k}\cdot\mathbf{p}}{m} + \frac{\hbar k^2}{2m} \tag{2.22}$$

is the excitation frequency of an electron-hole pair.

We note here that the four terms on the right-hand side of (2.21) govern the evolution of density-fluctuation excitations in the electron system. These terms mean as follows: First, the term (2.21a) describes free motions of the electron-hole pairs. The terms (2.21b) and (2.21c), on the other hand, stem from the Coulomb interaction. The former represents a mean field contribution linear in the density fluctuations, while the latter describes the effects of nonlinear coupling between the density fluctuations. Finally, the term (2.21d) accounts for the coupling between the density fluctuations and the external potential.

2.2.1 DENSITY–DENSITY RESPONSE FUNCTIONS

The dielectric function, $\varepsilon(\mathbf{k}, \omega)$, is defined in terms of the linear response relation between the external and induced potentials, $\Phi^{\text{ext}}(\mathbf{k}, \omega)$ and $\Phi^{\text{ind}}(\mathbf{k}, \omega)$, expressed as

$$\Phi^{\text{ind}}(\mathbf{k},\omega) = \left[\frac{1}{\varepsilon(\mathbf{k},\omega)} - 1\right]\Phi^{\text{ext}}(\mathbf{k},\omega). \tag{2.23}$$

The potential induced by the presence of the external potential is then given in terms of the induced density fluctuations,

$$\delta n_\sigma(\mathbf{k},\omega) = \int_{-\infty}^{\infty} dt \left\langle \sum_{\mathbf{p}} \rho_{\mathbf{pk}\sigma}(t) \right\rangle \exp(i\omega t) \tag{2.24a}$$

$$= \sum_\tau \chi_{\sigma\tau}(\mathbf{k},\omega) Z_\tau e \Phi_\tau^{\text{ext}}(\mathbf{k},\omega). \tag{2.24b}$$

In (2.24a), <···> means a statistical average over the states of the plasma. Equation (2.24b) defines the *density–density response functions*, $\chi_{\sigma\tau}(\mathbf{k}, \omega)$, between particles of σ and τ species.

The dielectric function defined by (2.23) is then calculated as

$$\frac{1}{\varepsilon(\mathbf{k},\omega)} = 1 + v(k)\sum_{\sigma,\tau} Z_\sigma Z_\tau \chi_{\sigma\tau}(\mathbf{k},\omega). \tag{2.25}$$

2.2.2 CORRELATIONS, RADIAL DISTRIBUTIONS, AND STATISTICAL THERMODYNAMICS

With the aid of the density–density response functions formulated in the preceding section and the fluctuation–dissipation theorem summarized in Appendix III, *the dynamic structure factors* $S_{\sigma\tau}(\mathbf{k}, \omega)$, *the static structure factors* $S_{\sigma\tau}(\mathbf{k})$, and *the radial distributions* $g_{\sigma\tau}(\mathbf{r})$ are calculated as

$$S_{\sigma\tau}(\mathbf{k},\omega) = -\frac{\hbar}{2\pi}\coth\left(\frac{\hbar\omega}{2k_B T}\right)\mathrm{Im}\,\chi_{\sigma\tau}(\mathbf{k},\omega), \tag{2.26}$$

$$S_{\sigma\tau}(\mathbf{k}) = \frac{1}{\sqrt{n_\sigma n_\tau}}\int_{-\infty}^{\infty} d\omega\, S_{\sigma\tau}(\mathbf{k},\omega), \tag{2.27}$$

$$g_{\sigma\tau}(\mathbf{r}) = 1 + \frac{1}{\sqrt{n_\sigma n_\tau}}\int_{-\infty}^{\infty} \frac{d\mathbf{k}}{(2\pi)^3}\left[S_{\sigma\tau}(\mathbf{k}) - \delta_{\sigma\tau}\right]\exp(i\mathbf{k}\cdot\mathbf{r}). \tag{2.28}$$

These functions describe the two-particle distributions in various versions and play the essential parts in formulating the equations of state, thermodynamic functions, and transport properties.

As for the thermodynamic functions in multi-ionic plasmas, the interaction energy per unit volume is given by

$$U_{\mathrm{int}} = \sum_{\sigma,\tau}\int_{-\infty}^{\infty}\frac{d\mathbf{k}}{(2\pi)^3}\frac{2\pi Z_\sigma Z_\tau e^2\sqrt{n_\sigma n_\tau}}{k^2}\left[S_{\sigma\tau}(\mathbf{k}) - \delta_{\sigma\tau}\right], \tag{2.29}$$

a counterpart to analogous quantity in (2.10).

The excess Helmholtz free energy per unit volume is then calculated with the coupling-constant integration of (2.29) as

$$F_{\mathrm{ex}} = \int_0^1 \frac{d\eta}{\eta} U_{\mathrm{int}}(\eta). \tag{2.30}$$

Here $U_{\mathrm{int}}(\eta)$ refers to the interaction energy (2.29) evaluated in a system where the strength of Coulomb coupling e^2 is replaced by ηe^2.

The excess pressure is given by

$$P_{\mathrm{ex}} = -\left(\frac{\partial F_{\mathrm{ex}}}{\partial V}\right)_{T,N_\sigma,N_\tau}, \tag{2.31}$$

where V is the volume and $N_\sigma = n_\sigma V$.

2.2.3 SPIN-DENSITY RESPONSE

The density–density response formalism of Sec. 2.2.1 may be transformed into a description of the spin-density responses for the electron system as follows: Let an external magnetic field, $H^{\text{ext}}(\mathbf{k}, \omega)\exp[i(\mathbf{k}\cdot\mathbf{r}-\omega t)]$, be applied to the electron system in an arbitrary direction, which will then induce spin-density fluctuations $\delta n_\sigma(\mathbf{k}, \omega)$. The *spin susceptibility*, $\chi^s(\mathbf{k}, \omega)$, is defined in accordance with

$$-\frac{g\mu_B}{2}\left[\delta n_{\uparrow}(\mathbf{k},\omega)-\delta n_{\downarrow}(\mathbf{k},\omega)\right]=\chi^s(\mathbf{k},\omega)H^{\text{ext}}(\mathbf{k},\omega), \tag{2.32}$$

where $g=2.0023$ is the g-factor and $\mu_B=e\hbar/2mc$ is the Bohr magneton. Since (2.24b) may be re-expressed in these circumstances as

$$\delta n_\sigma(\mathbf{k},\omega)=\frac{g\mu_B}{2}\sum_{\tau}\tau\chi_{\sigma\tau}(\mathbf{k},\omega)H^{\text{ext}}(\mathbf{k},\omega), \tag{2.33}$$

we find

$$\chi^s(\mathbf{k},\omega)=-\left(\frac{g\mu_B}{2}\right)^2\left[\chi_{\uparrow\uparrow}(\mathbf{k},\omega)+\chi_{\downarrow\downarrow}(\mathbf{k},\omega)-\chi_{\uparrow\downarrow}(\mathbf{k},\omega)-\chi_{\downarrow\uparrow}(\mathbf{k},\omega)\right]. \tag{2.34}$$

Spin-dependent, two-particle distribution functions can be calculated in terms of those spin-dependent response functions with the aid of the fluctuation–dissipation theorem (Appendix III) in the way analogous to (2.26)–(2.28).

2.2.4 THE HARTREE–FOCK APPROXIMATION

The density–density response functions of Sec. 2.2.1 may be calculated through a solution to the equation of motion (2.21), which governs the evolution of the density-fluctuation excitations in the electron system in the presence of the external Hamiltonian (2.18).

In the *Hartree–Fock approximation*, Coulomb-interaction terms (2.21b) and (2.21c) are ignored in the calculation of the response functions. In this approximation, the density responses are decoupled between different species and take a form,

$$\chi_{\sigma\tau}^{\text{HF}}(\mathbf{k},\omega)=\sum_{\rho}\delta_{\sigma\rho}\delta_{\tau\rho}\chi_{\rho}^{(0)}(\mathbf{k},\omega), \tag{2.35}$$

where the free-electron polarizability of the σ species is given by

$$\chi_{\sigma}^{(0)}(\mathbf{k},\omega)=\frac{1}{\hbar}\sum_{\mathbf{p}}\frac{1}{\omega-\omega_{\mathbf{pk}}+i0}\left[f_\sigma(\mathbf{p})-f_\sigma(\mathbf{p}+\hbar\mathbf{k})\right] \tag{2.36}$$

with the Fermi distribution

$$f_\sigma(\mathbf{p}) = \frac{1}{\exp\left[\left(\varepsilon_{\mathbf{p}} - \mu_\sigma\right)/k_B T\right] + 1}. \tag{2.37}$$

Here

$$\varepsilon_{\mathbf{p}} = \frac{p^2}{2m} \tag{2.38}$$

is the kinetic energy of an electron and μ_σ is the chemical potential for the free-electron system of the σ species, which is determined by the normalization,

$$\int \frac{d\mathbf{p}}{(2\pi\hbar)^3} f_\sigma(\mathbf{p}) = n_\sigma, \tag{2.39}$$

with n_σ denoting the number density of the spin-σ electrons.

In the treatment of a free-electron system at a finite temperature, it is useful to define the Fermi integrals,

$$I_\nu(\alpha) = \int_0^\infty dx \frac{x^\nu}{\exp(x-\alpha)+1}. \tag{2.40}$$

Mathematical properties of the Fermi integrals are summarized in Appendix IV. When the electron system is in the ground system ($T = 0$), the Fermi distribution takes a step function; Equation (2.36) evaluated in these circumstances is called the *Lindhard polarizability* (Lindhard, 1954; Pines & Nozières, 1966). In the classical limit of high temperature and low density, the Fermi distribution (2.36) approaches the Maxwell–Boltzmann form (1.29) and then (2.36) becomes the *Vlasov polarizability* (Vlasov, 1967; Ichimaru, 1973); we shall revisit this classical version in Sec. 3.3.

2.2.5 THE RANDOM-PHASE APPROXIMATION

The *random-phase approximation* (RPA) (Bohm & Pines, 1953), which we have introduced in Sec. 1.3.5, is an approach that goes beyond the Hartree–Fock approximation. In the RPA, one takes account of the Coulomb interaction, neglected in the Hartree–Fock approximation, through the mean field term (2.21b); the nonlinear coupling term (2.21c) between the density-fluctuation excitations remains neglected.

Since the mean field term is linear in the fluctuations, the RPA density–density response functions are calculated as

$$\chi_{\sigma\tau}^{\mathrm{RPA}}(\mathbf{k},\omega) = \sum_\rho \delta_{\sigma\rho}\delta_{\tau\rho}\chi_\rho^{(0)}(\mathbf{k},\omega) + \frac{Z_\sigma Z_\tau \chi_\rho^{(0)}(\mathbf{k},\omega)\chi_\tau^{(0)}(\mathbf{k},\omega)}{\varepsilon_0(\mathbf{k},\omega)}, \tag{2.41}$$

where

$$\varepsilon_0(\mathbf{k},\omega)=1-\sum_\sigma Z_\sigma^2 v(k)\chi_\sigma^{(0)}(\mathbf{k},\omega) \tag{2.42}$$

is the *RPA dielectric function.*

For an electron system in the ground state, the static (i.e., $\omega=0$) values of the Lindhard polarizability are evaluated as

$$\chi_\sigma^{(0)}(k,0)=-\frac{3n_\sigma m}{(\hbar k_{F\sigma})^2}\left\{\frac{1}{2}+\frac{k_{F\sigma}}{2k}\left(1-\frac{k^2}{4k_{F\sigma}^2}\right)\ln\left|\frac{k+2k_{F\sigma}}{k-2k_{F\sigma}}\right|\right\}, \tag{2.43}$$

where

$$k_{F\sigma}=\left(6\pi^2 n_\sigma\right)^{1/3} \tag{2.44}$$

is the *Fermi wavenumber* appropriate to the fermions of the σ species. In the limit of long wavelengths, the static RPA screening function, given by $\varepsilon_0(k,0)$, is then expressed as

$$\varepsilon_0(k,0)\to 1+\frac{k_{TF}^2}{k^2}\equiv 1+\frac{1}{k^2}\sum_\sigma \frac{6\pi n_\sigma (Z_\sigma e)^2}{(\hbar k_{F\sigma})^2/2m}, \tag{2.45}$$

which defines the *Thomas–Fermi screening parameter* k_{TF}.

For high-temperature and low-density classical plasmas, the static (i.e., $\omega=0$) values of the Vlasov polarizability are evaluated as

$$\chi_\sigma^{(0)}(k,0)=-\frac{n_\sigma}{k_B T}. \tag{2.46}$$

The static RPA screening function is therefore expressed as

$$\varepsilon_0(k,0)=1+\frac{k_D^2}{k^2}\equiv 1+\frac{1}{k^2}\sum_\sigma \frac{4\pi n_\sigma (Z_\sigma e)^2}{k_B T}, \tag{2.47}$$

which defines the *Debye–Hückel screening parameter* k_D, as in (1.32).

2.2.6 COLLECTIVE VERSUS INDIVIDUAL-PARTICLES ASPECTS OF FLUCTUATIONS

The collective versus individual-particles aspects of fluctuations described in Sec. 1.3.5 can be most succinctly described in terms of the *dielectric response function*, $\varepsilon(\mathbf{k},\omega)$. It is a linear response function as an externally applied potential $\Phi_{ext}(\mathbf{k},\omega)$ may induce a potential $\Phi_{ind}(\mathbf{k},\omega)$ introduced through (2.23); the resultant total potential field $\Phi_{tot}(\mathbf{k},\omega)$ ($=\Phi_{ext}(\mathbf{k},\omega)+\Phi_{ind}(\mathbf{k},\omega)$) may then be expressed as

$$\Phi_{\text{tot}}(\mathbf{k},\omega) = \Phi_{\text{ext}}(\mathbf{k},\omega) / \varepsilon(\mathbf{k},\omega). \tag{2.48}$$

The zeros of the dielectric response function, determined from $\varepsilon(\mathbf{k},\omega) = 0$ on the complex ω-plane, that is, $\omega = \omega_{\mathbf{k}} + i\gamma_{\mathbf{k}}$, give the frequency dispersion and the lifetime of the collective mode.

In an electron OCP, density fluctuations of individual electrons moving in trajectories, $\mathbf{r}_j(t) = \mathbf{r}_j + \mathbf{v}_j t$, are expressed as

$$\rho_j^{(0)}(\mathbf{k},\omega) = 2\pi \exp(-i\mathbf{k}\cdot\mathbf{r}_j)\delta(\omega - \mathbf{k}\cdot\mathbf{v}_j); \tag{2.49a}$$

those of the dressed electrons may then be given by

$$\rho_j^{(s)}(\mathbf{k},\omega) = \rho_j^{(0)}(\mathbf{k},\omega) / \varepsilon(\mathbf{k},\omega). \tag{2.49b}$$

In RPA, where collisions between dressed particles are negligible, the dynamic structure factor (2.26) may be evaluated by *superposition of the dressed test charges* as (Rostoker & Rosenbluth, 1960: Ichimaru, 1962)

$$S(\mathbf{k},\omega) = S^{(0)}(\mathbf{k},\omega) / \left|\varepsilon(\mathbf{k},\omega)\right|^2. \tag{2.50}$$

Here, the expression

$$S^{(0)}(\mathbf{k},\omega) = \int d\mathbf{v} f(\mathbf{v})\delta(\omega - \mathbf{k}\cdot\mathbf{v}) \tag{2.51a}$$

is applicable to a non-equilibrium situation such as a beam-plasma system, with $f(\mathbf{v})$ denoting the single-particle velocity distribution; the expression

$$S^{(0)}(\mathbf{k},\omega) = \frac{n}{k}\left(\frac{m}{2\pi k_B T}\right)^{1/2} \exp\left(-\frac{m\omega^2}{2k_B T k^2}\right) \tag{2.51b}$$

is applicable to a plasma in thermodynamic equilibrium at temperature T, where the Maxwellian (1.29) applies.

2.2.7 STRONG COUPLING EFFECTS

In the RPA of the previous sections, the strong exchange and Coulomb coupling effects represented by the nonlinear term (2.21c) of the fluctuations have not been taken into consideration. A theoretical method of going beyond such an RPA description and thereby accounting for the strong inter-particle correlation effects rigorously in the framework of the dielectric formulation has been provided by a *polarization potential approach* (Pines, 1963; Ichimaru, 2004b), which we shall summarize in this section.

We first note that the density responses stemming from the solution to (2.21a–d) may be written in this approach as

$$\delta n_\sigma(\mathbf{k},\omega) = \chi_\sigma^{(0)}(\mathbf{k},\omega)\left\{ Z_\sigma e\Phi_\sigma^{\text{ext}}(\mathbf{k},\omega) + \sum_\tau Z_\sigma Z_\tau v(k)\left[1 - G_{\sigma\tau}(\mathbf{k},\omega)\right]\delta n_\tau(\mathbf{k},\omega)\right\}. \quad (2.52)$$

This relation thus implies that the induced density fluctuations, $\delta n_\tau(\mathbf{k}, \omega)$, of the particles of the τ species may produce an effective potential field of the strength

$$Z_\sigma e\Phi_{\sigma\tau}^{\text{pol}}(\mathbf{k},\omega) = Z_\sigma Z_\tau v(k)\left[1 - G_{\sigma\tau}(\mathbf{k},\omega)\right]\delta n_\tau(\mathbf{k},\omega), \quad (2.53)$$

which acts on the particles of the σ species; $\Phi_{\sigma\tau}^{\text{pol}}(\mathbf{k},\omega)$ in (2.53) may be looked upon as a polarization potential in these circumstances. Such a polarization potential generally differs from the RPA mean field, $Z_\sigma Z_\tau v(k)\delta n_\tau(\mathbf{k}, \omega)$, since the exchange and Coulomb correlations between the particles in effect modify the potentials due to the induced space charges. The differences are measured by the *dynamic local-field correlations* $G_{\sigma\tau}(\mathbf{k}, \omega)$ in (2.53), which originates from the nonlinear term (2.21c) in the equation of motion for the density-fluctuation excitations. The density–density response functions may then be calculated explicitly from the solution to (2.52) in accordance with (2.24b).

In certain cases of the condensed plasma problems, one adopts an approximation whereby the dynamic local-field corrections are assumed to be independent of the frequency variables and are replaced by their static evaluations,

$$G_{\sigma\tau}(\mathbf{k},\omega) \to G_{\sigma\tau}(\mathbf{k}) \equiv G_{\sigma\tau}(\mathbf{k},\omega = 0). \quad (2.54)$$

The functions $G_{\sigma\tau}(\mathbf{k})$ are called the *static local-field corrections*, and a theoretical scheme of treating the strong coupling effects via $G_{\sigma\tau}(\mathbf{k})$ will be referred to as the *static local-field correction approximation.*

Let us now revisit the spin-density responses in such a static local-field approximation for an electron system in the paramagnetic state, in which

$$n_{-} = n_\downarrow = \frac{n}{2}. \quad (2.55)$$

Since the system is isotropic, it is appropriate to define the local-field corrections for parallel and antiparallel spins, $G_p(k)$ and $G_a(k)$, and the total free-electron polarizability $\chi_0(k, \omega)$ via

$$G_p(k) = G_{--}(k) = G_{\downarrow\downarrow}(k), \quad (2.56a)$$

$$G_a(k) = G_{-\downarrow}(k) = G_{\downarrow -}(k), \quad (2.56b)$$

$$\chi_0(k,\omega) = 2\chi_{-}^{(0)}(k,\omega) = 2\chi_\downarrow^{(0)}(k,\omega) \quad (2.57)$$

(e.g., Ichimaru, Iyetomi, & Tanaka, 1987).

The dielectric function of (2.25) and the spin susceptibility of (2.32) are then given by

$$\varepsilon(k,\omega)=1-\frac{v(k)\chi_0(k,\omega)}{1+v(k)G(k)\chi_0(k,\omega)}, \tag{2.58}$$

$$\chi^s(k,\omega)=-\left(\frac{g\mu_B}{2}\right)^2\frac{\chi_0(k,\omega)}{1+J(k)\chi_0(k,\omega)}. \tag{2.59}$$

Here,

$$G(k)=\frac{1}{2}\left[G_p(k)+G_a(k)\right], \tag{2.60}$$

$$J(k)=\frac{v(k)}{2}\left[G_p(k)-G_a(k)\right]. \tag{2.61}$$

In the limit of long wavelengths, these fluctuations are related to the *isothermal compressibility* κ_T and the static spin susceptibility χ_P via the thermodynamic sum rules treated in Appendix III. Since one defines

$$\kappa_T=\frac{1}{n}\left(\frac{\partial n}{\partial P}\right)_T, \tag{2.62}$$

$$\chi_P=\left(\frac{\partial M}{\partial B}\right)_T, \tag{2.63}$$

with P, M, and B denoting the pressure, the density of spin magnetization, and the strength of the magnetic field, respectively, the *compressibility sum rule* and the *spin-susceptibility sum rule* read

$$\lim_{k\to 0} v(k)G(k)=-\frac{1}{\chi_0(0,0)}-\frac{1}{n^2\kappa_T}, \tag{2.64}$$

$$\lim_{k\to 0} J(k)=-\frac{1}{\chi_0(0,0)}-\left(\frac{g\mu_B}{2}\right)^2\frac{1}{\chi_P}. \tag{2.65}$$

These sum-rule relations provide a crucial linkage between the dielectric formulation and the *Landau theory of Fermi liquids* (Landau, 1956, 1957; Pines & Nozières, 1966). The extension to the regime of finite wavelengths offers an essential ingredient of the density-functional theory, which we shall consider in the next section.

2.3 DENSITY-FUNCTIONAL THEORY

Consider a system of N_σ electrons contained in a box of volume V under the influence of external potentials $\phi_\sigma(\mathbf{r})$, where σ denotes the spin coordinate. The field operators, which are constructed through coherent superposition of the creation and annihilation operators of Sec. 2.1.2, may be expressed as

$$\psi_\sigma^\dagger(\mathbf{r}) = \frac{1}{\sqrt{V}} \sum_{\mathbf{p}} c_{\mathbf{p}\sigma}^\dagger \exp\left(-\frac{i}{\hbar} \mathbf{p} \cdot \mathbf{r} \right), \tag{2.66a}$$

$$\psi_\sigma(\mathbf{r}) = \frac{1}{\sqrt{V}} \sum_{\mathbf{p}} c_{\mathbf{p}\sigma} \exp\left(\frac{i}{\hbar} \mathbf{p} \cdot \mathbf{r} \right). \tag{2.66b}$$

The density operators are given by

$$n_\sigma(\mathbf{r}) = \psi_\sigma^\dagger(\mathbf{r}) \psi_\sigma(\mathbf{r}). \tag{2.67a}$$

We assume that the system of electrons under consideration has a unique, nondegenerate ground state Ψ. Clearly, then, Ψ is a unique functional of $\phi_\sigma(\mathbf{r})$ and, therefore, so are the expectation values of the electron densities,

$$n_\sigma(\mathbf{r}) = (\Psi, n_\sigma(\mathbf{r})\Psi). \tag{2.67b}$$

The important conclusion derived by Hohenberg and Kohn (1964) is that $\phi_\sigma(\mathbf{r})$ and Ψ, in turn, are uniquely determined by the knowledge of the density distributions, $n_\sigma(\mathbf{r})$. This conclusion lays the theoretical foundation to the *density-functional theory* (e.g., Kohn & Vashishta, 1983; Callaway & March, 1984), which we shall summarize in this section.

2.3.1 KOHN–SHAM SELF-CONSISTENT EQUATIONS

The total Hamiltonian operator of the system may be written as a sum of the kinetic, internal energy, and external contributions,

$$H_{\text{tot}} = H_K + H_{\text{int}} + H_{\text{ext}}, \tag{2.68}$$

where

$$H_K = \frac{\hbar^2}{2m} \sum_\sigma \int d\mathbf{r} \nabla \psi_\sigma^\dagger(\mathbf{r}) \cdot \nabla \psi_\sigma(\mathbf{r}), \tag{2.69a}$$

$$H_{\mathrm{int}} = \frac{1}{2}\sum_{\sigma,\tau}\int d\mathbf{r}d\mathbf{r}'\frac{Z_\sigma Z_\tau e^2}{|\mathbf{r}-\mathbf{r}'|}\psi_\sigma^\dagger(\mathbf{r})\psi_\tau^\dagger(\mathbf{r}')\psi_\tau(\mathbf{r}')\psi_\sigma(\mathbf{r}), \tag{2.69b}$$

$$H_{\mathrm{ext}} = \sum_\sigma\int d\mathbf{r}Z_\sigma e\phi_\sigma\psi_\sigma^\dagger(\mathbf{r})\psi_\sigma(\mathbf{r}). \tag{2.69c}$$

Expectation values, designated by angular brackets, of these contributions and therefore the total Hamiltonian are the functionals of the densities, $n_\sigma(\mathbf{r})$. We write, in particular,

$$\begin{aligned}\langle H_K + H_{\mathrm{int}}\rangle &= E_K[n_\sigma(\mathbf{r})]\\ &+\frac{1}{2}\sum_{\sigma,\tau}\int d\mathbf{r}d\mathbf{r}'\frac{Z_\sigma Z_\tau e^2}{|\mathbf{r}-\mathbf{r}'|}n_\sigma(\mathbf{r})n_\tau(\mathbf{r}') + E_{\mathrm{xc}}[n_\sigma(\mathbf{r})],\end{aligned} \tag{2.70}$$

where $E_K[n_\sigma(\mathbf{r})]$ refers to the kinetic energy of a *non-interacting* electron system with the densities $n_\sigma(\mathbf{r})$ in the ground state, the second is the Hartree interaction term, and the remainder, $E_{\mathrm{xc}}[n_\sigma(\mathbf{r})]$, is called the *exchange and correlation energy*. The expectation value of the total Hamiltonian is thus expressed as

$$\begin{aligned}E_{\phi 0}[n_\sigma(\mathbf{r})] &= E_K[n_\sigma(\mathbf{r})] + \frac{1}{2}\sum_{\sigma,\tau}\int d\mathbf{r}d\mathbf{r}'\frac{Z_\sigma Z_\tau e^2}{|\mathbf{r}-\mathbf{r}'|}n_\sigma(\mathbf{r})n_\tau(\mathbf{r}')\\ &+E_{\mathrm{xc}}[n_\sigma(\mathbf{r})] + \sum_\sigma\int d\mathbf{r}Z_\sigma e\phi_\sigma(\mathbf{r})n_\sigma(\mathbf{r}).\end{aligned} \tag{2.71}$$

Its dependence on the external potentials $\phi_\sigma(\mathbf{r})$ has been denoted explicitly in this equation.

In the ground state, the expectation values (2.71) take on the minimal values with respect to variations of the densities around themselves, subject to conservation of the numbers of the particles,

$$\int d\mathbf{r}n_\sigma(\mathbf{r}) = N_\sigma. \tag{2.72}$$

The resultant Euler equations ensuing from the density-functional derivatives (see Appendix V) of (2.71) are

$$\frac{\delta E_K[n_\sigma]}{\delta n_\sigma(\mathbf{r})} + Z_\sigma e\phi_\sigma^{\mathrm{tot}}(\mathbf{r}) + v_\sigma^{\mathrm{xc}}(\mathbf{r}) - \mu_\sigma = 0. \tag{2.73}$$

Here,

$$\phi_\sigma^{\mathrm{tot}}(\mathbf{r}) = \phi_\sigma(\mathbf{r}) + \sum_\tau\int d\mathbf{r}'\frac{Z_\tau e}{|\mathbf{r}-\mathbf{r}'|}n_\tau(\mathbf{r}') \tag{2.74}$$

refer to the total classical potentials,

$$v_\sigma^{\text{xc}}(\mathbf{r}) \equiv \frac{\delta E_{\text{xc}}[n_\sigma]}{\delta n_\sigma(\mathbf{r})} \tag{2.75}$$

are the *exchange–correlation potentials* defined via the functional derivatives, and μ_σ are the Lagrange parameters associated with the subsidiary conditions (2.72).

The *Kohn–Sham self-consistent equations* for the single-particle wave functions $\psi_i(\mathbf{r})$ are derived in terms of those potentials as (Kohn & Sham, 1965)

$$\left[-\frac{\hbar^2}{2m}\nabla^2 + v_\sigma^{\text{eff}}(\mathbf{r})\right]\psi_i(\mathbf{r}) = \varepsilon_i\psi_i(\mathbf{r}), \tag{2.76}$$

where

$$v_\sigma^{\text{eff}}(\mathbf{r}) = Z_\sigma e\phi_\sigma^{\text{tot}}(\mathbf{r}) + v_\sigma^{\text{xc}}(\mathbf{r}). \tag{2.77}$$

The self-consistency is enforced via

$$n_\sigma(\mathbf{r}) = \sum_{i=1}^{N_\sigma}\left|\psi_i(\mathbf{r})\right|^2, \tag{2.78}$$

where the sum is to be carried out over the N_σ lowest occupied eigenstates.

Analytic theories accounting for the thermodynamic properties of dense plasmas are concerned with the derivation of relevant expressions for the local-field corrections, $G_{\sigma\tau}(k)$, as explained in Sec. 2.2.7. The local-field corrections can be formulated in the density-functional theory extended to the finite temperatures by Mermin (1965).

2.3.2 THERMODYNAMIC POTENTIALS

The thermodynamic potentials of interacting inhomogeneous plasmas in external potentials $\phi_\sigma(\mathbf{r})$ may be expressed as

$$\begin{aligned}\Omega_{\phi_\sigma}\left[n_\sigma(\mathbf{r})\right] = F_0\left[n_\sigma(\mathbf{r})\right] + \frac{1}{2}\sum_{\sigma,\tau}\int d\mathbf{r}d\mathbf{r}'\frac{Z_\sigma Z_\tau e^2}{|\mathbf{r}-\mathbf{r}'|}n_\sigma(\mathbf{r})n_\tau(\mathbf{r}') \\ + F_{\text{xc}}\left[n_\sigma(\mathbf{r})\right] + \sum_\sigma\int d\mathbf{r}\left\{Z_\sigma e\phi_\sigma(\mathbf{r}) - \mu_\sigma\right\}n_\sigma(\mathbf{r}).\end{aligned} \tag{2.79}$$

Here, $F_0[n_\sigma(\mathbf{r})]$ denotes the free energy functional of a free-particle system with density distributions $n_\sigma(\mathbf{r})$, and $F_{\text{xc}}[n_\sigma(\mathbf{r})]$ refers to the remaining exchange–correlation free energy functional for the interacting system. Assuming that the densities are expressed in the form

$$n_\sigma(\mathbf{r}) = n_\sigma + \delta n_\sigma(\mathbf{r}), \tag{2.80a}$$

with $n_\sigma = N_\sigma / V$, $|\delta n_\sigma(\mathbf{r})| / n_\sigma \ll 1$, and

$$\int d\mathbf{r} \delta n_\sigma(\mathbf{r}) = 0, \tag{2.80b}$$

one finds the Euler equations for minimization of the thermodynamic potential (2.79) with respect to the density variations $\delta n_\sigma(\mathbf{r})$. That is,

$$\frac{\delta F_0[n_\sigma]}{\delta n_\sigma(\mathbf{r})} + v_\sigma^{\mathrm{xc}}(\mathbf{r}) + Z_\sigma e \phi_\sigma^{\mathrm{tot}}(\mathbf{r}) = 0, \tag{2.81}$$

where

$$v_\sigma^{\mathrm{xc}}(\mathbf{r}) \equiv \frac{\delta F_{\mathrm{xc}}[n_\sigma]}{\delta n_\sigma(\mathbf{r})} \tag{2.82}$$

are the exchange–correlation potentials at finite temperatures.

To the lowest order in $\delta n_\sigma(\mathbf{r})$, the first two terms in (2.81) should be linear in $\delta n_\sigma(\mathbf{r})$ on account of (2.80b), so that one writes

$$\sum_\tau \int d\mathbf{r}' \left[K_{\sigma\tau}^{(0)}(\mathbf{r}-\mathbf{r}'; n_\sigma) + K_{\sigma\tau}^{\mathrm{xc}}(\mathbf{r}-\mathbf{r}'; n_\sigma) + \frac{Z_\sigma Z_\tau e^2}{|\mathbf{r}-\mathbf{r}'|} \right] \delta n_\tau(\mathbf{r}') \\ + Z_\sigma e \phi_\sigma^{\mathrm{ext}}(\mathbf{r}) = 0. \tag{2.83}$$

The kernels, $K_{\sigma\tau}^{(0)}(\mathbf{r})$ and $K_{\sigma\tau}^{\mathrm{xc}}(\mathbf{r})$, introduced in (2.83), are the second density-functional derivatives of $F_0[n_\sigma(\mathbf{r})]$ and $F_{\mathrm{xc}}[n_\sigma(\mathbf{r})]$ around $n_\sigma(\mathbf{r}) = n_\sigma$; they depend only on $\mathbf{r}-\mathbf{r}'$ and n_σ, the average number densities. The spatial Fourier transformation of (2.83) yields

$$\sum_\tau \left[\tilde{K}_{\sigma\tau}^{(0)}(\mathbf{k}; n_\sigma) + \tilde{K}_{\sigma\tau}^{\mathrm{xc}}(\mathbf{k}; n_\sigma) + Z_\sigma Z_\tau v(k) \right] \delta\tilde{n}_\tau(\mathbf{k}) + Z_\sigma e \Phi_\sigma^{\mathrm{ext}}(\mathbf{k}, 0) = 0. \tag{2.84}$$

Here, for example,

$$\tilde{K}_{\sigma\tau}^{(0)}(\mathbf{k}; n_\sigma) = \int d\mathbf{r} K_{\sigma\tau}^{(0)}(\mathbf{r}; n_\sigma) \exp(-i\mathbf{k}\cdot\mathbf{r}). \tag{2.85}$$

Equation (2.84) is a relation for the static linear response in the plasma. Direct density-functional calculations and comparison with (2.35) and (2.52) yield

$$\sum_\rho \tilde{K}_{\sigma\rho}^{(0)}(\mathbf{k}; n_\sigma) \chi_{\rho\tau}^{\mathrm{HF}}(\mathbf{k}, 0) = -\delta_{\sigma\tau}, \tag{2.86}$$

$$\tilde{K}_{\sigma\tau}^{\mathrm{xc}}(\mathbf{k}; n_\sigma) = -Z_\sigma Z_\tau v(k) G_{\sigma\tau}(\mathbf{k}), \tag{2.87}$$

where $G_{\sigma\tau}(\mathbf{k})$ are the static local-field corrections defined in (2.54). Hence, we may find (2.84) as equivalent to (2.52) in the static local-field approximation.

2.4 COMPUTER SIMULATION METHODS

There exist a number of computer simulation methods that aim at studying the interparticle correlations and thermodynamic properties of plasmas based on the fundamental principles of statistical physics. This section summarizes some of those approaches.

2.4.1 MONTE CARLO APPROACHES

A *Monte Carlo* (MC) method is any method making use of random numbers to solve a problem (James, 1980). The power of the MC method lies basically in its ability to carry through multidimensional integrations through the techniques of importance sampling with improved accuracy as well as with increased capacity of modern computers (e.g., Binder, 1979, 1992).

In the Metropolis algorithm (Metropolis et al., 1953), one works with a statistical ensemble at a constant temperature, volume, and number of particles, that is, the *canonical ensemble.* Monte Carlo steps (configurations) are generated by random displacements of particles in the many-particle system under study. Configurations so created will be accepted or rejected with the probability of acceptance:

$$P = \begin{cases} \exp(-\Delta E / k_B T), & \text{if} \quad \Delta E > 0, \\ 1, & \text{if} \quad \Delta E \leq 0, \end{cases} \tag{2.88}$$

where ΔE denotes the energy increment created by the displacements. A Markov chain representing the canonical ensemble is thereby generated, with its thermalization usually monitored through evaluation of the internal energy. The probability (2.88) may thus lead the system to one with a canonical distribution at temperature T. In their pioneering work, Brush, Sahlin, & Teller (1966) performed numerical experiments on strongly coupled OCPs by such an MC method.

2.4.2 MOLECULAR DYNAMICS SIMULATIONS

In the method of *molecular dynamics* (MD), pioneered by Alder and Wainwright (1959), the classical equations of motion for a system of interacting particles are solved by integration in discrete time steps; equilibrium properties are determined from time averages taken over a sufficiently long time interval (e.g., Cicconti, Frenkel, & McDonald, 1987; Yonezawa, 1992).

In the MD simulation, one thus works with the *microcanonical ensemble,* in which the total energy, volume, and number of particles are kept constant; the MD methods are therefore deterministic in principle. There exist provisions (Andersen, 1980; Hoover et al., 1980; Nosé & Klein, 1983) in the framework of the MD method that can simulate *isothermal* and/or *isobaric* ensembles.

In the approach proposed by Car and Parrinello (1985), dynamic aspects of an MD method are accommodated in the treatment of simulated annealing for a quantum many-body system through a density-functional theory described in Sec. 2.3. In these connections, we take it essential that the self-consistency in the formulation and application of the Kohn–Sham equations (2.76) be strictly maintained. These requirements include accounting of the dynamic polarization potentials (2.53) and/or the exchange–correlation potentials (2.75) in the simulations.

2.4.3 OTHER APPROACHES

Quantum many-body problems may likewise be approached through a number of other computer techniques (e.g., Ceperley & Kalos, 1979; Binder, 1979, 1992; Yonezawa, 1992).

In the *variational MC method* (McMillan, 1965), one calculates an expectation value,

$$E_T = \frac{\int d\mathbf{R}\psi_T(\mathbf{R})H(\mathbf{R})\psi_T(\mathbf{R})}{\int d\mathbf{R}\left|\psi_T(\mathbf{R})\right|^2}, \tag{2.89}$$

of the N-particle Hamiltonian, H(**R**), with a trial wave function, $\psi_T(R)$, through multidimensional MC integrations, where

$$\mathbf{R} = (\mathbf{r}_1, \cdots, \mathbf{r}_N) \tag{2.90}$$

refers to the coordinates of the N particles.

Variational trial functions for a system of fermions are sometimes chosen in a form of a Jastrow (1955) type,

$$\psi_J(\mathbf{R}) = \psi_0(\mathbf{R})\exp\left[-\frac{1}{2}\sum_{i<j} u(r_{ij})\right], \tag{2.91}$$

where $\psi_0(\mathbf{R})$ is the ideal Fermi gas wave function and $u(r_{ij})$ corresponds to a variational "pseudopotential" accounting for inter-particle correlation with $r_{ij} = |\mathbf{r}_i - \mathbf{r}_j|$. A solution to the Schrödinger equation,

$$H(\mathbf{R})\psi(\mathbf{R}) = E\psi(\mathbf{R}), \tag{2.92}$$

is obtained for the ground state by minimizing the energy E_T of (2.89) through variations of the trial functions.

In *Green's Function Monte Carlo* (GFMC) method (e.g., Ceperley & Kalos, 1979), one seeks an integral formulation of the Schrödinger equation and considers Green's function for the left-hand side of (2.92), that is,

$$\left[-\frac{\hbar^2}{2m}\nabla_{\mathbf{R}}^2 + V(\mathbf{R}) + V_0\right]\mathbf{G}(\mathbf{R},\mathbf{R}_0) = \delta(\mathbf{R}-\mathbf{R}_0). \tag{2.93}$$

Here the Hamiltonian has been split into the kinetic and potential energy contributions and $-V_0$ refers to a minimum of $V(\mathbf{R})$. It is then possible to devise an MC method, in a general sense of a random sampling algorithm, that produces populations drawn from the successive $\psi^{(n)}(\mathbf{R})$ generated by

$$\psi^{(n+1)}(\mathbf{R}) = (E+V_0)\int d\mathbf{R}'\mathbf{G}(\mathbf{R},\mathbf{R}')\psi^{(n)}(\mathbf{R}'). \tag{2.94}$$

As the iterations converge, a solution to the Schrödinger equation may be obtained.

Ceperley and Alder (1980), in particular, have calculated the ground-state energy of the electron plasma by the GFMC method. Progress in the computer simulation studies on the properties of hydrogen and helium under extreme conditions has been extensively reviewed (McMahon et al., 2012).

3

SCATTERING OF ELECTROMAGNETIC WAVES

Scattering of electromagnetic waves is a useful way by which one studies fluctuations and correlations in the plasmas. To elucidate the principles of such approaches, we begin with an evaluation of the cross-sections of individual and correlated charged particles against scattering of electromagnetic waves. We then proceed to look into the radar backscattering experiments that demonstrate those collective processes in the ionosphere.

3.1 SCATTERING BY INDIVIDUAL PARTICLES

Consider an electron with electric charge -e and mass m located in the plane-wave electromagnetic field, whose electric field is expressed as

$$\mathbf{E}(\mathbf{r},t) = \mathbf{E}_1 \cos(\mathbf{k}_1 \cdot \mathbf{r} - \omega_1 t). \tag{3.1}$$

The quantities ω_1 and k_1 satisfy $\omega_1/k_1 = c$, where $c = 2.9978 \times 10^{10}$ cm/s is the light velocity in vacuum. The electron and its dipole moment $\mathbf{d}$ obey the equation of motion,

$$\ddot{\mathbf{d}} = -e\ddot{\mathbf{r}} = \frac{e^2}{m}\mathbf{E}.$$

Since the rate of energy emitted by an electric dipole is $2\left|\ddot{\mathbf{d}}\right|^2 / 3c^3$, the emission rate from the electron subjected to the electromagnetic field (3.1) is calculated as

$$\frac{2}{3c^3}\frac{e^4}{m^2}E^2 = \frac{8\pi}{3}\left(\frac{e^2}{mc^2}\right)^2\left(\frac{E^2}{4\pi}\right)c. \tag{3.2}$$

3.1.1 CROSS-SECTION OF THOMSON SCATTERING

The incident electromagnetic wave carries a total energy density of $E^2/4\pi$, since the electric field and the magnetic field carry the same amount of energy density $E^2/8\pi$. The energy flux of the incident wave is $(E^2/4\pi)c$, and so we find from (3.2) the scattering cross-section of an electron is

$$\sigma_T = \frac{8\pi}{3}\left(\frac{e^2}{mc^2}\right)^2 = 6.653\times 10^{25}\ \text{cm}^2. \tag{3.3}$$

The quantity is called the *cross-section of Thomson scattering.*

As may be clear in (3.3), the scattering cross-section is inversely proportional to the square of the mass. In an ordinary gaseous plasma, the mass of an ion is at least 1837 times that of an electron; thus, the cross-sections of electrons are predominant, as far as scattering of electromagnetic wave is concerned.

If the electromagnetic waves scattered by individual electrons do not interfere with each other, that is, if the correlation effects between electrons are negligible, the total cross-section of the plasma against scattering of the electromagnetic wave is proportional to the number of electrons in the scattering volume. In these circumstances, one can deduce the electron density from a measurement of the total intensity of scattered waves.

3.1.2 DOPPLER EFFECT

We have thus found that the total cross-section of an electron against scattering of the electromagnetic wave is given by the cross-section (3.3) of Thomson scattering. Generally, however, the frequency of the scattered wave differs from that of the incident wave due to the Doppler effect associated with the motion of electrons.

Let $\mathbf{v}$ be the velocity of an electron in the laboratory system (see Figure 3.1). As we find by substituting the orbital $\mathbf{r} = \mathbf{r}(t = 0) + \mathbf{v}t$ of the electron motion in (3.1), the electron perceives $\omega' = \omega_1 - \mathbf{k}_1\cdot\mathbf{v}$ as the frequency of the incident wave. Such a motional change of the frequency is called the *Doppler effect*. The electron with the velocity $\mathbf{v}$ thus emits an electromagnetic wave with frequency ω' on its own frame of reference. Then, we, in the frame of reference fixed to the laboratory system, observe radiation emitted in the direction of the wave vector $\mathbf{k}_2$. With the aid of Doppler's relation once again, we find the frequency of the scattered wave at $\omega_2 = \omega' + \mathbf{k}_2\cdot\mathbf{v}$. Consequently, a frequency shift takes place between the incident and scattered waves with a magnitude,

$$\omega_2 - \omega_1 = \mathbf{v}\cdot(\mathbf{k}_2 - \mathbf{k}_1). \tag{3.4}$$

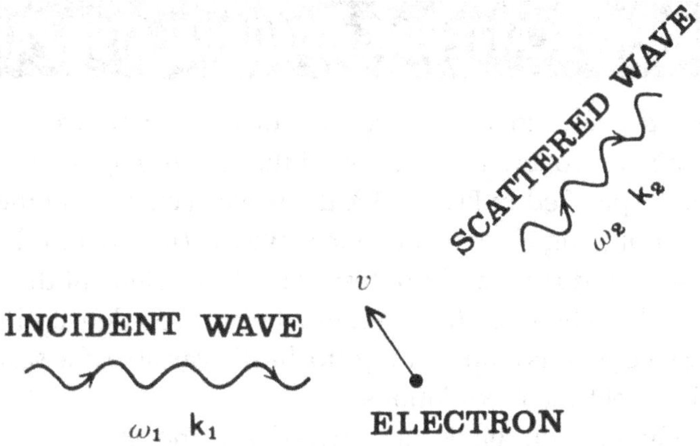

FIGURE 3.1 Scattering of electromagnetic wave by an electron.

If the incident wave is monochromatic with the frequency ω_1, the frequency ω_2 of the scattered wave exhibits a characteristic spread corresponding to the velocity distribution of the electrons. For a Maxwellian distribution (1.29) with temperature T, the frequency spectrum of the scattered wave thus takes the form as shown in Figure 3.2 where

$$\omega = \omega_2 - \omega_1, \tag{3.5}$$

$$\mathbf{k} = \mathbf{k}_2 - \mathbf{k}_1, \tag{3.6}$$

and $M = m$. In these circumstances, the electron temperature can be determined from the spread in the frequency spectrum of the scattered waves. Equations (3.5) and (3.6) correspond to the conservation laws of energy and momentum in the event of such scattering.

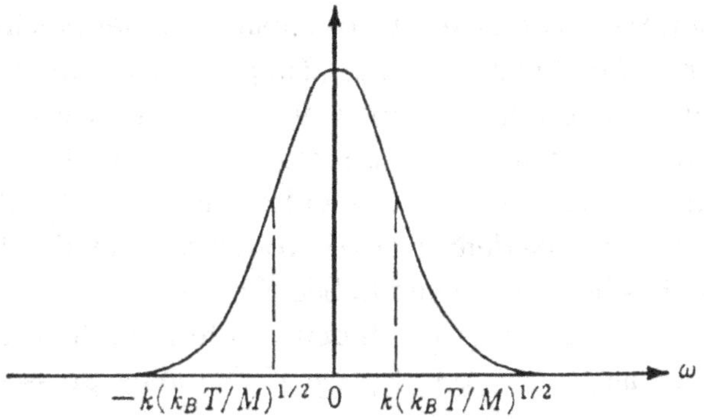

FIGURE 3.2 Frequency spectrum of individual-particle-like density fluctuations.

3.2 INCOHERENT SCATTERING BY CORRELATED PARTICLES

Let us now proceed to consider the effects of correlations between electrons on the cross-sections against incoherent scattering of the electromagnetic waves.

We designate, as specified in Figure 3.3, the wave vector **k** and the frequency ω of the incoming and outgoing electromagnetic waves as $(\mathbf{k}_1, \omega_1)$ and $(\mathbf{k}_2, \omega_2)$. The intensity of the wave scattered from an infinitesimal volume element $d\mathbf{r}_1$ around $\mathbf{r}_1$ is proportional to $\rho(\mathbf{r}_1, t)\, d\mathbf{r}_1$, where $\rho(\mathbf{r}, t)$ represents the electron density in the plasma at **r** and t – cf. (1.22). Since $\rho(\mathbf{r}, t)$ is a fluctuating stochastic variable, the scattered wave also carries stochastic amplitude modulations as

(amplitude modulation on the wave scattered at $\mathbf{r}_1$ and t_1)

$$\sim \int_V d\mathbf{r}\, \rho(\mathbf{r}_1, t_1) \exp\left[i(\mathbf{k}_1 - \mathbf{k}_2)\cdot \mathbf{r}_1 - i(\omega_1 - \omega_2)t_1\right] \tag{3.7}$$

We then carry out a spectral decomposition for the mean-square average of the stochastically fluctuating fields (3.7):

$$\begin{aligned} \left\langle |\text{amplitude}|^2 \right\rangle \sim \int_V & d\mathbf{r}_1 d\mathbf{r}_2 \left\langle \rho(\mathbf{r}_1, t_1)\rho(\mathbf{r}_2, t_2)\right\rangle \\ & \exp\left[i(\mathbf{k}_1 - \mathbf{k}_2)\cdot(\mathbf{r}_1 - \mathbf{r}_2) - i(\omega_1 - \omega_2)(t_1 - t_2)\right] \end{aligned} \tag{3.8}$$

Assuming that the plasma in the scattering volume V is uniform and stationary, we may take the space-and-time correlation function $\left\langle \rho(\mathbf{r}_1, t_1)\rho(\mathbf{r}_2, t_2)\right\rangle$ of the density fluctuations as a function of only the variables, $\mathbf{r}_1 - \mathbf{r}_2$ and $t_1 - t_2$. In terms of the Fourier components, we then express

$$\left\langle \rho(\mathbf{r}_1, t_1)\rho(\mathbf{r}_2, t_2)\right\rangle = \frac{1}{V}\sum_{\mathbf{k}} \int d\omega\; S(\mathbf{k}, \omega) \exp\left[i\mathbf{k}\cdot(\mathbf{r}_1 - \mathbf{r}_2) - i\omega(t_1 - t_2)\right]. \tag{3.9}$$

The spectral function $S(\mathbf{k}, \omega)$ of the electron-density fluctuations introduced in (3.9) is the *dynamic structure factor*; it is the same as (2.6) presented earlier.

Consideration of (3.7) combined with (3.9) may convince us that the differential cross-sections of the plasma against scattering of the electromagnetic wave is expressed in a form proportional to $S(\mathbf{k}, \omega)$ with the aid of (3.5) and (3.6). The constant of proportionality may then be determined through the way that the cross-section of Thomson scattering has been evaluated in Sec. 3.1.1.

The differential cross-section for incoherent scattering into a solid angle do and a frequency interval dω, as portrayed in Figure 3.3, is thus expressed in the formula as

$$\frac{d^2Q}{do\, d\omega} = \frac{3V}{8\pi}\sigma_T \left(1 - \frac{1}{2}\sin^2\theta\right) S(\mathbf{k}, \omega). \tag{3.10}$$

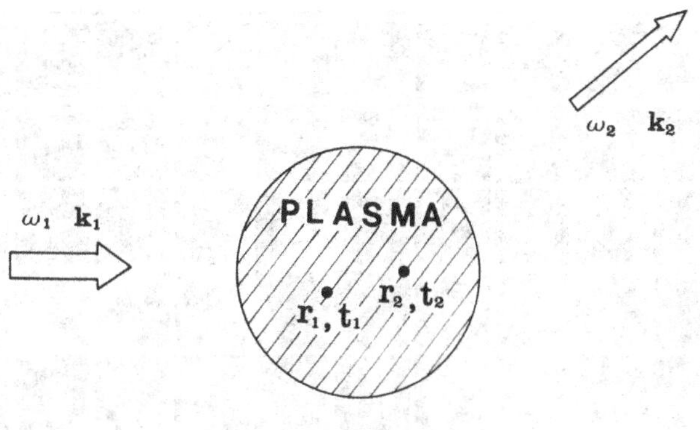

FIGURE 3.3 Incoherent scattering of electromagnetic waves.

Here $\mathbf{k} = \mathbf{k}_2 - \mathbf{k}_1$, $\omega = \omega_2 - \omega_1$, θ is the angle between the incident and scattered waves, and we have averaged over directions of polarization of both waves (e.g., Ichimaru, 1973). Scattering of electromagnetic waves thus provides a unique way of monitoring electron–electron density correlations in the plasma.

3.3 RADAR BACKSCATTERING FROM THE IONOSPHERE

Around the year 1960, a series of experiments were performed to probe the state of the ionospheric plasmas through backscattering of radar pulses. It was originally conceived that the time-dependence of the scattered signal would reveal directly the electron density and temperature profiles as functions of the altitude. Unexpectedly, however, the measured results revealed collective features in the ionospheric plasmas. It was particularly notable that the observed spectrum of scattered waves carried a strong influence of the ionic motions despite the fact that the electrons were the scatterers. Let us therefore study the features of inter-particle correlations specific in electrons-and-ions, two-component plasmas.

3.3.1 OBSERVATIONS BY BOWLES

For a first example of scattering experiments that demonstrate correlation effects in plasmas, let us take up radar backscattering from the ionosphere, as carried out by Bowles of the National Bureau of Standards (NBS) (Bowles, 1958, 1961); it is the experiments particularly mentioned in the first paragraph of the Rosenbluth and Rostoker (1962) scattering paper.

The F-layer of the ionosphere, consisting of electrons and ions (mostly oxygen), extends from 200 to 500 km in altitude; the average number densities are around 10^5–10^6 cm^{-3} at temperatures about 1500 K. Assuming the maximum electron density to be around 2×10^6 cm^{-3}, we find the corresponding plasma frequency at 12.7 MHz. This is the critical frequency for the propagation of electromagnetic waves in the plasma; waves with frequencies less than that will be reflected by the F-layer.

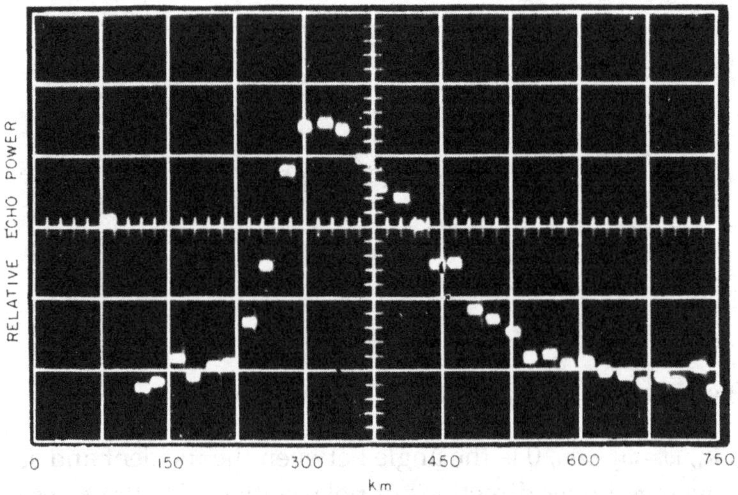

FIGURE 3.4 Radar-backscattering measurement of the ionospheric electron distribution as a function of the altitude [1959 Feb. 27, 7:40 p.m. – Illinois local time]. After Bowles (1961).

An electromagnetic wave with a frequency exceeding the critical frequency, on the other hand, propagates through the F-layer; only a tiny fraction is scattered by the electrons and sent back to the earth's surface.

The radar backscattering experiments that Bowles carried out in 1958 utilize radar pulses at frequency 42.92 MHz, pulse-width 120 μs, repetition frequency 25~40 per second, and peak power 1 MW. Since the radar frequency is far greater than the plasma frequency at 13 MHz, the ionosphere is transparent to those radar pulses and backscatters them at strengths proportional to the local densities of electrons. The density profile so observed is shown in Figure 3.4.

We expect the backscattered waves would be broadened, on top of a 9 kHz spread in the radar frequencies, by a width of 82 kHz with the thermal motion of the electrons (at $T = 1500$ K) as in (2.51b) as well as in Figure 3.2. To detect such a spectral broadening, the receiver's bandwidth was fixed at 9 kHz.

When the central frequency of the receiver was set at the outgoing radar frequency, a signal with a maximum strength was obtained. When it was shifted by 15 kHz relative to that radar frequency, however, virtually *no signals* were observed. This casts an *enigma* since it might mean *no broadening* caused by the scatterers.

3.3.2 OBSERVATIONS BY PINEO, KRAFT, AND BRISCOE

To look into the features of broadening more closely, Pineo, Kraft, and Briscoe (1960) of MIT two years later performed analogous experiments, in which, however, the outgoing radar frequency was raised to 440 MHz, almost an order of magnitude greater than that used by Bowles. Frequency spectrum of the backscattered waves has now become detectable, as in Figure 3.5; we here find that the broadening does actually take place correspondingly, however, to the *thermal motion of the ions.*

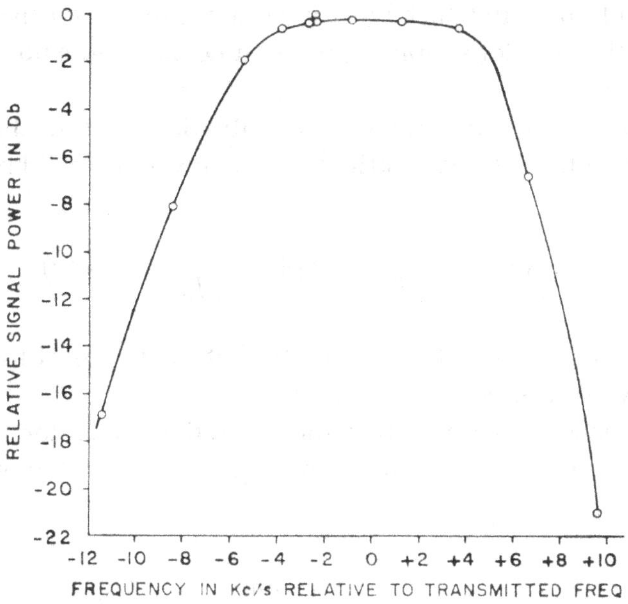

FIGURE 3.5 Spectral distribution of backscattered wave from ionosphere (approximately 300 km in altitude) (Pineo, Kraft, & Briscoe, 1960).

Now, in these experiments, we recognize that the wave number k (= $2k_1$) relevant to backscattering is much smaller than the Debye wave number k_D of (1.32), meaning that we are in the *collective regime* where effects of the correlation are predominant.

In the present case of electron-and-ion, two-component plasmas, the features of the collective versus individual-particles aspects of the fluctuations have to be substantially altered from those with the electron one-component plasma (OCP) described in Secs. 1.3.4 and 1.3.5.

First, the presence of "dressed" ions should be noted. Coulomb potential around an ion is screened by co-moving clouds of electrons at half the strength and the remaining half stems from repelled ions. The observed spectrum in Figure 3.5 may be interpreted as coming from scattering by those electrons co-moving with ions. These features will further be explained in Sec. 3.4.2.

We must also note the appearance of a new collective mode, called the ion-acoustic waves that may affect features of the electron-density fluctuations. These will be treated in Sec. 3.4.4.

3.4 COLLECTIVE PHENOMENA IN ELECTRON-AND-ION PLASMAS

In the present case of ionospheric plasmas, which are of two components, collective oscillations likewise consist of two modes, *optical* and *acoustic*. For the treatment of those collective modes, we thus extend the dielectric formulations of Sec. 2.3 to two-component plasmas, in which we use the subscript "e" for the electrons and the subscript "i" for the ions; for simplicity, we assume the charge number Z of an ion to be unity, and $n_e = n_i = n$.

In this section, we are concerned with a situation close to the thermodynamic equilibrium, so that the velocity distributions are given by the Maxwellian,

$$f_\sigma(\mathbf{v}) = \left(\frac{m_\sigma}{2\pi k_B T_\sigma}\right)^{1/2} \exp\left(-\frac{m_\sigma v^2}{2k_B T_\sigma}\right) \quad (\sigma = e, i). \tag{3.11}$$

In so doing, we are assuming the possibility that the temperatures of the electrons and the ions may be different.

In these connections, we note that the relaxation times for Maxwellization of electrons and of ions and for temperature equality are in the approximate ratios (Ichimaru, 1973),

$$\tau_{ee} : \tau_{ii} : \tau_{ei} \sim 1 : (m_i / m_e)^{1/2} : m_i / m_e. \tag{3.12}$$

Since we may take $m_i/m_e \gg 1$ generally for plasmas, use of Maxwellians with unequal temperatures may be looked upon as reasonable; we find such an unequal temperature plasma in the glow discharges, for instance.

3.4.1 DIELECTRIC RESPONSE FUNCTION

The dielectric response function in the random-phase approximation (RPA) for the Maxwellian plasma is then calculated as

$$\varepsilon(\mathbf{k},\omega) = 1 - \sum_\sigma \frac{\omega_\sigma^2}{k^2} \int d\mathbf{v} \frac{1}{\mathbf{k}\cdot\mathbf{v} - \omega - i\eta} \mathbf{k} \cdot \frac{\partial f_\sigma(\mathbf{v})}{\partial \mathbf{v}} \tag{3.13}$$

with

$$\omega_\sigma = \left(4\pi n e^2 / m_\sigma\right)^{1/2} \quad (\sigma = e, i). \tag{3.14}$$

Here the positive infinitesimal η serves to assure the adiabatic turning on of the disturbance and thereby to guarantee a *causal response* of the system; we let $\eta \to +0$ eventually (Ichimaru, 1973).

We now substitute (3.11) in (3.13) to obtain

$$\varepsilon(\mathbf{k},\omega) = 1 + \frac{k_e^2}{k^2} W\left(\frac{\omega}{k\sqrt{k_B T_e / m_e}}\right) + \frac{k_i^2}{k^2} W\left(\frac{\omega}{k\sqrt{k_B T_i / m_i}}\right) \tag{3.15}$$

with

$$k_\sigma = \left(4\pi ne^2 / k_B T_\sigma\right)^{1/2} \quad (\sigma = e, i). \tag{3.16}$$

Here the W function is the error function of a complex argument (Fried & Conte, 1961; Ichimaru, 1973):

$$W(Z) = 1 - Z\exp\left(-\frac{Z^2}{2}\right)\int_0^Z dy \exp\left(\frac{y^2}{2}\right) + i\left(\frac{\pi}{2}\right)^{1/2} Z\exp\left(-\frac{Z^2}{2}\right). \tag{3.17}$$

For $|Z| < 1$, it can be expressed in a convergent series,

$$W(Z) = i\left(\frac{\pi}{2}\right)^{1/2} Z\exp\left(-\frac{Z^2}{2}\right) + 1 - Z^2 + \frac{Z^4}{3} - \cdots + \frac{(-1)^{n+1} Z^{2n+2}}{(2n+1)!!} \cdots \tag{3.18}$$

where

$$(2n+1)!! = (2n+1)(2n-1)\cdots 3 \cdot 1.$$

For large Z, we have an asymptotic series

$$W(Z) = i\left(\frac{\pi}{2}\right)^{1/2} Z\exp\left(-\frac{Z^2}{2}\right) - \frac{1}{Z^2} - \frac{3}{Z^4} - \cdots - \frac{(2n-1)!!}{Z^{2n}} - \cdots. \tag{3.19}$$

3.4.2 DRESSED PARTICLES

Let us now revisit (2.27) in Sec. 2.2.1, and evaluate in particular the electronic static structure factor, $S_e(k)$, in a two-component plasma, given by and calculated as (Ichimaru, 1962)

$$S_e(k) \equiv \sum_{\tau=e,i} S_{e\tau}(k) = \frac{k^2 + k_i^2}{k^2 + k_e^2 + k_i^2}. \tag{3.20a}$$

Assuming $k_e = k_i$, one finds $S_e = 1/2$ in the long-wavelength domain satisfying $k^2 \ll k_e^2$. It is noteworthy that this value differs markedly from that evaluated for the electron OCP (e.g., Ichimaru, 1973):

$$S_e(k) = \frac{k^2}{k^2 + k_e^2}, \tag{3.20b}$$

which vanishes as $k \to 0$.

Interpreted physically, the limiting value 1/2 with (3.20a) stems from the contribution of those electrons participating in the screening of individual ions. Since the static screening effects of electrons and ions are the same, the "dress" of an ion is a 50–50 mixture of electrons and other ions.

Those electrons, which form the correlated screening clouds of ions, carry the total cross-section as much as a half of the Thomson scattering cross-section of individual electrons even in the long-wavelength domain of $k^2 \ll k_e^2$. These may offer the explanation to Figure 3.5.

3.4.3 ION-ACOUSTIC WAVES

The observational results exhibited in Figure 3.5 may also suggest a trace of another collective mode, *ion-acoustic waves*, in the two-component plasmas. We thus investigate the collective modes in the long-wavelength regime, $k^2 \ll k_\sigma^2$, for electron–ion plasmas, setting the solution to $\varepsilon(\mathbf{k},\omega)=0$ as $\omega=\omega_\mathbf{k}+\gamma_\mathbf{k}$.

In the high-frequency regime such that

$$|\omega| \gg k\sqrt{k_B T_e / m_e} \gg k\sqrt{k_B T_i / m_i}, \tag{3.21}$$

the dielectric response function (3.15) may be expressed with the large Z expansion (3.19) for both electrons and ions; we thus find

$$\omega_\mathbf{k} = \omega_e\left[1+\frac{3}{2}\left(\frac{k}{k_e}\right)^2+\cdots\right], \tag{3.22a}$$

$$\frac{\gamma_\mathbf{k}}{\omega_\mathbf{k}} = -\left(\frac{\pi}{8}\right)^{1/2}\left(\frac{k}{k_e}\right)^3 \exp\left(-\frac{k_e^2}{2k^2}\right). \tag{3.22b}$$

This is basically a space-charge wave of electrons, in a uniform, positive-charge background of ions; it represents the optical mode of the plasma oscillation.

In the intermediate-frequency regime such that

$$k\sqrt{k_B T_e / m_e} \gg |\omega| \gg k\sqrt{k_B T_i / m_i}, \tag{3.23}$$

the dielectric response function (3.15) may be expressed with the small Z expansion (3.18) for electrons and with the large Z expansion (3.19) for ions; we thus find

$$\omega = \omega_k\left\{\pm 1 - i\left(\frac{\pi}{8}\right)^{1/2}\left[\left(\frac{m_e}{m_i}\right)^{1/2}+\left(\frac{T_e}{T_i}\right)^{3/2}\exp\left(-\frac{T_e}{2T_i}-\frac{3}{2}\right)\right]\right\}. \tag{3.24}$$

with

$$\omega_k = \left[\left(k_B T_e + 3k_B T_i\right)/m_i\right]^{1/2} k. \tag{3.25}$$

This is the acoustic mode of plasma oscillations representing the density waves of electron-screened ions. These acoustic waves in the ionosphere, however, are strongly damped since the temperature of the electrons is almost equal to that of the ions, or $T_e \simeq T_i$; hence, only a mild hump has been observed in Figure 3.5 as a trace suggesting of this corrective mode.

If, however, one considers the plasma with $T_e \gg T_i$, then the ion-acoustic waves, $\omega_k = sk$, with the sound velocity,

$$s = \sqrt{k_B T_e / m_i}, \tag{3.26}$$

become well-defined, relatively undamped oscillations with the decay rate γ_k, given by

$$\frac{\gamma_k}{\omega_k} = \sqrt{\frac{\pi m_e}{8 m_i}}. \tag{3.27}$$

Numerically, $|\gamma_k/\omega_k| \approx 0.015$ for a hydrogen plasma. They are the density waves of ions interacting mutually via electron-screened, short-ranged Coulomb forces, analogous to the phonons in ordinary materials.

Finally, on the basis of the foregoing considerations, let us review the experiments by Pineo et al. in Sec. 3.3.2. For a hydrogen plasma with $T_e \approx 1500$ K, we find $s \approx 3.5 \times 10^5$ cm/s from (3.26). Since $k = 2k_1 = 0.18$ cm^{-1} and $k_e = 5.3$ cm^{-1} from (3.16), we confirm $k^2 \ll k_e^2$ and the frequency of the ion-acoustic wave is approximately 10 kHz. In the case of the ionospheric plasma where oxygen ions are involved, the ion-acoustic frequency may take on a value smaller than that. At any rate, those ion-acoustic waves suffer severe damping because $T_e \simeq T_i$ in the ionosphere and act only to produce weak shoulder structures as observed in Figure 3.5.

3.5 PLASMA CRITICAL OPALESCENCE

In the RPA, the dynamic structure factor of the electrons for the electron–ion plasma may be simply obtained by superposing the fields due to the dressed particles (Ichimaru, Pines, & Rostoker, 1962); the result is

$$S(\mathbf{k},\omega) = \frac{N}{k} f_e(\omega/k) \left| \frac{1 + 4\pi\alpha_i(\mathbf{k},\omega)}{\varepsilon(\mathbf{k},\omega)} \right|^2 + \frac{N}{k} f_i(\omega/k) \left| \frac{4\pi\alpha_e(\mathbf{k},\omega)}{\varepsilon(\mathbf{k},\omega)} \right|^2, \tag{3.28}$$

where

$$\alpha_\sigma(k,\omega) = \lim_{\eta \to 0} -\frac{ne^2}{m_\sigma k^2} \int_{-\infty}^{\infty} dv \frac{k\left[df_\sigma(v)/dv\right]}{kv - \omega - i|\eta|} \quad (\sigma = e, i) \tag{3.29}$$

are the polarizabilities of electrons (e) and ions (i); $f_\sigma(v)$ are the one-dimensional normalized velocity distribution functions in the directions of $\mathbf{k}$.

Drift motion of the electrons relative to the ions acts to excite the ion-acoustic waves. If one passes to a sufficiently large value of the drift velocity V_d, the ion-acoustic waves become unstable; the boundary between growing and damped waves is specified by

$$\mathrm{Im}\,\varepsilon(\mathbf{k},\omega_k) = 0, \tag{3.30}$$

where ω_k is determined by

$$\mathrm{Re}\,\varepsilon(\mathbf{k},\omega_k) = 0. \tag{3.31}$$

Since $S(\mathbf{k},\omega)$ in (3.28) is proportional to $1/|\varepsilon(\mathbf{k},\omega)|^2$, one finds a contribution from the immediate vicinity of ω_k, which is

$$S_{\mathrm{res}}(\mathbf{k}) \propto 1/\mathrm{Im}\,\varepsilon(\mathbf{k},\omega_k). \tag{3.32}$$

For the case of marginal stability, defined by (3.30), $S_{\mathrm{res}}(\mathbf{k})$ obviously diverges.

We have carried out an explicit evaluation of $S_{\mathrm{res}}(\mathbf{k})$ for the case that $T_e \gg T_i$, such that the waves close to $k=0$ are the first to grow as one increases V_d. The result is (Ichimaru, Pines, & Rostoker, 1962)

$$S_{\mathrm{res}}(\mathbf{k}) = \frac{1}{2}\frac{\sqrt{m_e/m_i}}{(V_c - V_d\cos\chi)/V_e + (k/k_e)^2(V_1/V_e)}, \tag{3.33}$$

where $V_e = \sqrt{k_B T_e/m_e}$, $V_c \cong s$, χ is the angle between $\mathbf{k}$ and $\mathbf{V}_d$, and

$$\frac{V_1}{V_e} = \frac{1}{2}\left\{\left(\frac{T_e}{T_i}\right)^{5/2}\exp\left[-\frac{(T_e/T_i)+3}{2}\right] - \left(\frac{m_e}{m_i}\right)^{1/2}\right\}. \tag{3.34}$$

The result (3.33) is valid for $k^2 \ll k_e^2$.

We remark that (3.33) is identical in analytical form to the results obtained for the critical fluctuations in the vicinity of a liquid-gas phase transition, or the critical opalescence (Landau & Lifshitz, 1969).

3.6 OBSERVATION OF PLASMA WAVES IN WARM DENSE MATTER

Rader backscattering from the ionosphere described in Sec. 3.3 and the plasma critical opalescence treated in Sec. 3.5 are concerned with observation of the collective phenomena in plasmas.

X-ray Thomson scattering techniques have likewise been employed for observation of collective modes in warm dense matter (Glenzer et al., 2007). The measurements were performed in solid-density beryllium target that had been heated isochorically with a broadband X-ray source into a state of dense plasma with weakly degenerate electrons. The collective scattering regime of the plasma, that is, $k \ll k_D$ in (3.10), was then approached through forward scattering (i.e., $\theta \ll 1$) of the narrow-band chlorine Ly-α X-ray line at 2.96 keV.

The characteristic peak associated with the collective plasma oscillations (e.g., Ichimaru, 1973) has thereby been observed in agreement with the theoretical responses, in which the collisional effects have been appropriately taken into account.

4

CHARGED PARTICLES OR X-RAYS INJECTED IN PLASMAS

The fundamental properties of the plasmas can be investigated by means of the scattering of incident X-ray or particle beams. The spectrum of electron-density fluctuations such as the plasma oscillations can be monitored through the energy-loss spectroscopy in the transmission experiments. It is also possible to probe the microscopic features of electron correlation through the injection of molecular-ionic beams into metal. We here begin with a consideration of the properties of electron plasmas in metal, which can be detected by those scattering methods.

4.1 CHARACTERISTIC ENERGY-LOSS SPECTROSCOPY

The X-ray scattering spectroscopy is schematically illustrated in Figure 4.1. The characteristic X-rays of a metal are usually employed as the incident X-rays. For example, the $K\alpha_2$ line of copper has the energy 8.052 keV; the $K\alpha_2$ line of copper has the energy 8.031 keV (1,544 Å in wavelength) and the natural width 3.5 eV. Transmitting such X-ray into a thin metallic plate, one measures the spectral distribution of the scattered X-ray with an angle θ relative to the incident direction. The differential cross-section of scattering is given by (3.10); the energy and momentum of the incident X-ray, $\hbar\omega_1$ and $\hbar\mathbf{k}_1$, and those of the scattered X-ray, $\hbar\omega_2$ and $\hbar\mathbf{k}_2$, are connected via (3.5) and (3.6).

The difference $\hbar\omega$ between the incident and scattered X-rays coincides with the energy of density-fluctuation excitations in the electron plasma. The plasma oscillation is a typical example of such collective excitations. Its quantum, called a *plasmon*, has the energy

$$\hbar\omega_p = 47.1 r_s^{-3/2}\,[\text{eV}], \tag{4.1}$$

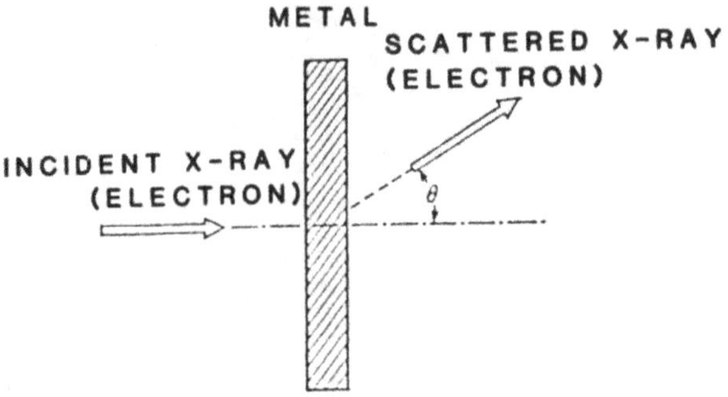

FIGURE 4.1 X-ray (electron) transmission-scattering spectroscopy.

where r_s is the density parameter introduced by (1.4).

Although this expression does not take into account any effective-mass correction to the metallic electrons or the dielectric constant of the ionic lattice, we may approximately estimate $\hbar\omega_p = 3.2 - 16.7$ eV for electrons at metallic densities (r_s=2–6). The numerical examples cited above indicate $\hbar\omega_1 \gg \hbar\omega_p$, so that we may take $\hbar\omega_1 \simeq \hbar\omega_2$. Hence, for $\theta \ll 1$,

$$k = (\omega_1 / c)\theta. \tag{4.2}$$

In these transmission-scattering experiments, one can alternatively use a beam of mono-energetic electrons (electron energy-loss spectroscopy) instead of the X-ray. In this case, a metal foil with a thickness of around 10^3 Å is used; its differential cross-section of scattering takes a form analogous to (3.10), i.e.,

$$\frac{d^2Q}{do d\omega} = \frac{Vm^2}{4\pi^2\hbar^4}\left(\frac{E_1}{E_2}\right)^{1/2}\left(\frac{4\pi e^2}{k^2}\right)^2 S(\mathbf{k},\omega) \tag{4.3}$$

(e.g., Pines, 1963). Denoting by $\mathbf{p}_1$, E_1, $\mathbf{p}_2$, and E_2 the momenta and energies of the incident and scattered electrons, we have relations corresponding to (3.5) and (3.6) as

$$\hbar\omega = E_2 - E_1, \tag{4.4}$$

$$\hbar\mathbf{k} = \mathbf{p}_2 - \mathbf{p}_1. \tag{4.5}$$

These relations connect between the energy and momentum of the incident and the excitation energy $\hbar\omega$ and momentum $\hbar\mathbf{k}$ of the density fluctuations.

In an actual experiment, the energy range of 20–300 keV is used for E_1; hence, as in the case of X-ray scattering, one can take $E_1 \simeq E_2$. For $\theta \ll 1$, one thus has

$$k = \frac{1}{\hbar}\sqrt{2mE_1}\theta. \tag{4.6}$$

The energy resolution of the scattered particles, depending on the experiment, is usually 0.1–1 eV.

4.2 PLASMON DISPERSION

Figure 4.2 shows an example of the energy-loss spectrum measured in an electron transmission experiment (Gibbons et al., 1976). The sample is aluminum foil with thickness 1100 Å. As we have noted earlier in Sec. 1.2.2, aluminum has $r_s = 2.1$ and $\hbar\omega_p = 15.5$ eV by (4.1). The peak energy is observed to increase with the wave number; the dispersion relation of the plasma oscillation can be determined experimentally through such an observation. One can likewise estimate the decay rate of plasma oscillation from the half width of the peak structure.

Theoretically, as we have argued in Sec. 2.2.6, the zeros of the dielectric response function, determined from $\varepsilon(\mathbf{k},\omega) = 0$ on the complex ω-plane, that is, $\omega = \omega_{\mathbf{k}} + i\gamma_{\mathbf{k}}$, give the frequency dispersion and the lifetime of the collective mode. In the static local-field correction approximation of Sec. 2.2.7, the dielectric function for the paramagnetic electrons has been expressed as Eq. (2.58), where the free-electron polarizability is given by

$$\chi_0(k,\omega) = \frac{2}{\hbar}\sum_p \frac{f(\mathbf{p}) - f(\mathbf{p}+\hbar\mathbf{k})}{\omega - \mathbf{k}\cdot\mathbf{p}/m - \hbar k^2/2m + i0}. \tag{4.7}$$

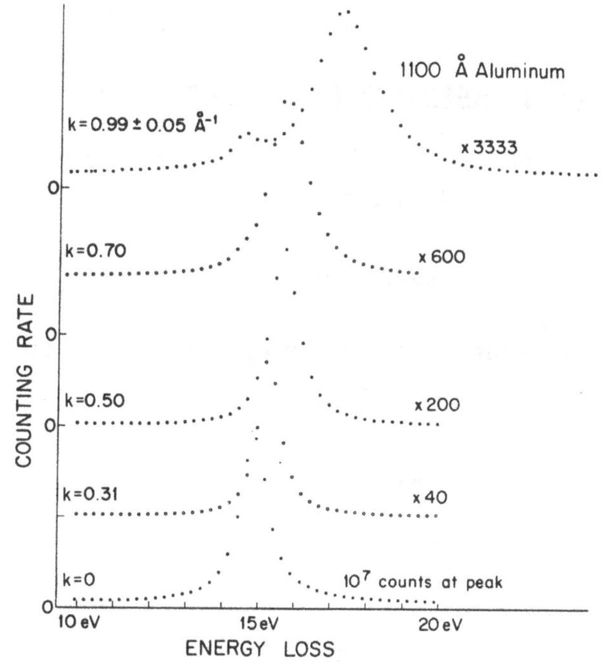

FIGURE 4.2 Electron energy-loss spectra of aluminum at various momentum transfers (Gibbons et al., 1976).

Here the momentum distribution function f(**p**) takes on values

$$f(\mathbf{p})=\begin{cases}1 & (p\le \hbar k_F)\\ 0 & (p>\hbar k_F)\end{cases} \tag{4.8}$$

representing the Fermi-sphere distribution with the Fermi wavenumber $k_F=(6\pi^2 n)^{1/3}$ for the metallic electrons; the electrons assume the maximum velocity $v_F=\hbar k_F/m$ on the Fermi sphere.

In the random-phase approximation (RPA) of Sec. 2.2.5, one ignores the strong-coupling effects such as the local-field corrections in (2.58). Assuming $|\omega/k|\gg v_F$, we carry out high-frequency expansion of the RPA dielectric function to obtain

$$\mathrm{Re}\,\varepsilon(k,\omega)=1-\frac{\omega_p^2}{\omega^2}-\frac{8\pi e^2}{m\omega^4}\sum_{\mathbf{p}} f(\mathbf{p})\left[3\left(\frac{\mathbf{k}\cdot\mathbf{p}}{m}\right)^2+\frac{\hbar^2 k^4}{4m^2}+\cdots\right]. \tag{4.9}$$

Retaining up to the terms proportional to k^2 in the long-wavelength domain ($k^2 \ll k_F^2$), we find

$$\omega_k^2=\omega_p^2+\frac{3}{5}v_F^2 k^2 \tag{4.10}$$

as a solution to Re $\varepsilon(k,\omega)=0$.

4.2.1 PLASMON DISPERSION COEFFICIENT

We define the *plasmon dispersion coefficient* α by expressing the measured plasmon energy as a function of the scattering angle θ as

$$\hbar\omega(\theta)=\hbar\omega_p+2\alpha E_1\theta^2. \tag{4.11}$$

Since the plasmon wave number and scattering angle are related via (4.6), this equation implies

$$\hbar\omega_k=\hbar\omega_p+\alpha\frac{\hbar^2}{m}k^2. \tag{4.12}$$

In the RPA, one thus has

$$\alpha_{\mathrm{RPA}}=\frac{3}{5}\frac{E_F}{\hbar\omega_p}=0.637 r_s^{-1/2} \tag{4.13}$$

in view of the dispersion relation (4.10).

4.2.2 MEASURED VALUES

The measured values of α for various metals are summarized in Figure 4.3 on the basis of the data compiled in *Excitations of Plasmons and Interband Transitions by Electrons* (Raether, 1980). The data are scattered rather widely partly because of different experimenters and experimental uncertainties.

Nevertheless, the observed results seem to indicate consistent deviations from α_{RPA}, which obviously widen as r_s increases. Thus, the strong-coupling effects beyond the RPA have been demonstrated through these experimental data.

4.2.3 THEORETICAL ESTIMATES

A number of theoretical attempts have been advanced to account for those deviations (Utsumi & Ichimaru, 1981).

An approach takes account of the long-wavelength limit of the static local-field correction as

$$\lim_{k \to 0} G(k) = \gamma_0(r_s)(k / k_F)^2, \tag{4.14}$$

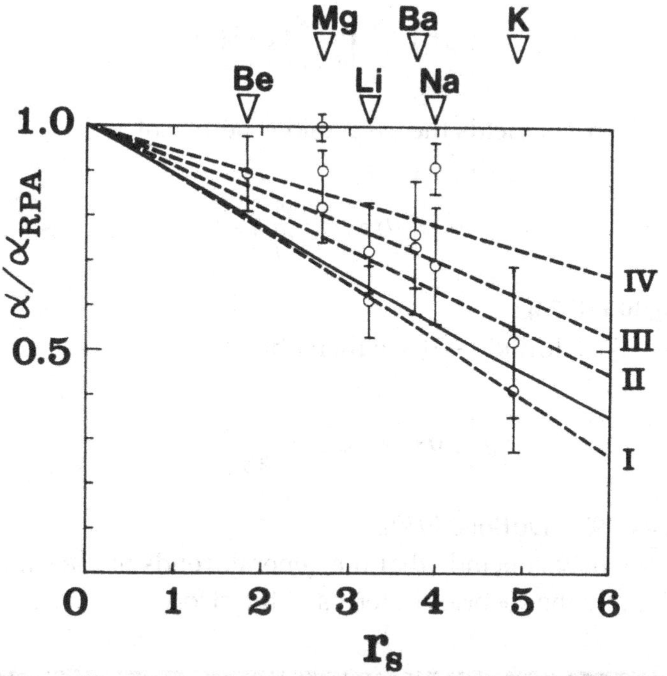

FIGURE 4.3 Plasmon dispersion coefficient α divided by RPA value α_{RPA} vs. r_s in various approximations (Utsumi & Ichimaru, 1981). Solid curve corresponds to the result of Utsumi & Ichimaru (1981). Dashed curves I-IV correspond to α_0, α_{TW}, α_∞, and α_{NP} in Equations (4.15), (4.16), (4.19), and (4.20), respectively. The experimental values with open circles and error bars are taken from (Raether, 1980).

and evaluates with the aid of the compressibility sum-rule (2.64) as

$$\alpha_0 = \alpha_{\text{RPA}} - \frac{\hbar\omega_p}{4E_F}\gamma_0(r_s) \tag{4.15}$$

(Singwi et al., 1968).

Obviously, this evaluation depends on r_s. Then, taking the Hartree–Fock limit of $r_s \to 0$ for $\gamma_0(r_s)$, Toigo and Woodruff (1971) proposed

$$\alpha_{\text{TW}} = \alpha_{\text{RPA}} - \frac{\hbar\omega_p}{16E_F}. \tag{4.16}$$

The plasma oscillations in metallic electrons being high-frequency phenomena, one may rather take the high-frequency and long-wavelength limit of the dynamic local-field corrections in (2.52), that is, $\lim_{\omega\to\infty} G(k,\omega) \equiv G_\infty(k)$ with

$$\lim_{k\to 0} G_\infty(k) = \gamma_\infty(r_s)(k/k_F)^2, \tag{4.17}$$

where

$$\gamma_\infty(r_s) = \frac{1}{5}\int_0^\infty \frac{dk}{k_F}[1 - S(k)]. \tag{4.18}$$

Use of (4.17) in (2.58) then yields the dispersion coefficient

$$\alpha_\infty = \alpha_{\text{RPA}} - \frac{\hbar\omega_p}{4E_F}\gamma_\infty(r_s) \tag{4.19}$$

(Pathak & Vashishta, 1973).

In the Hartree–Fock limit, $r_s \to 0$, we then obtain

$$\alpha_{\text{NP}} = \alpha_{\text{RPA}} - \frac{3\hbar\omega_p}{80E_F} \tag{4.20}$$

(Nozières & Pines, 1958; DuBois, 1959).

In summary, we may conclude that the general trends of the experiments are collectively described by the theories in terms of the strong Coulomb-coupling effects.

4.3 STOPPING POWER AND WAKE POTENTIAL

Let us now turn to consider energetic ions injected externally into metallic electrons. Those ions interact with the electrons and thereby probe the correlation properties in the electron plasmas.

4.3.1 INDUCED DENSITY VARIATIONS

To begin, we calculate the external potential that an injected ion with an electric charge Ze and mass M traveling at a velocity $\mathbf{u}$ exerts onto the degenerate electron plasma. We assume u to be much greater than the Fermi velocity v_F, that is,

$$u \gg v_F. \tag{4.21a}$$

For a proton, this implies

$$\frac{1}{2}Mu^2 \gg 92.1 r_s^{-2} \text{ (keV)}. \tag{4.21b}$$

The injected charge produces a potential of external disturbance in the notation of Sec. 2.2.1 as

$$\Phi^{\text{ext}}(\mathbf{k},\omega) = 8\pi^2 \frac{Ze}{k^2}\delta(\omega - \mathbf{k}\cdot\mathbf{u}). \tag{4.22}$$

The injected ion induces the electron-density fluctuations $\delta\rho(\mathbf{k},\omega)$ in the plasma according to (2.24). The space-time distribution of the induced electron-density fluctuations is obtained by carrying out the inverse Fourier transformation of $\delta\rho(\mathbf{k},\omega)$ as

$$\rho_{\text{ind}}(\mathbf{r},t) = Z\int \frac{d\mathbf{k}}{(2\pi)^3}\left[1 - \frac{1}{\varepsilon(\mathbf{k},\mathbf{k}\cdot\mathbf{u})}\right]\exp\left[i\mathbf{k}\cdot(\mathbf{r}-\mathbf{u}t)\right]. \tag{4.23}$$

The condition (4.21a) leads to the expansion (4.9) of the dielectric response function, of which we retain the leading terms. Adding to it the imaginary part, we express

$$\varepsilon(\mathbf{k},\omega) = 1 - \frac{\omega_p^2}{\omega^2} + i\eta\frac{\omega}{|\omega|}, \tag{4.24}$$

where η is a positive infinitesimal.

We calculate (4.23) by choosing the z-axis in the $\mathbf{u}$ direction. The integrations with respect to k_x and k_y in (4.23) then produce $\delta(x)$ and $\delta(y)$.

To carry out the k_z integration, we close the contour of integration by an infinite semicircle with $\text{Im}\, k_z > 0$ on the complex k_z plane when $z - ut > 0$ (see Figure 4.4). The integral pertaining to (4.23) vanishes by virtue of Cauchy's theorem because no roots of $\varepsilon(\mathbf{k}, k_z u) = 0$ exist on the upper half of the complex plane for (4.24).

For $z - ut < 0$, on the other hand, we close the contour of integration by an infinite semicircle with $\text{Im}\, k_z < 0$. In this case, the residues at the two roots of $\varepsilon(\mathbf{k}, k_z u) = 0$ in the lower half plane contributes to the integration as shown in Figure 4.4.

As a result, we find

$$\rho_{\text{ind}}(\mathbf{r},t) = -\frac{Z}{\lambda_p}\delta(x)\delta(y)\sin\left(\frac{z-ut}{\lambda_p}\right)\theta(ut-z). \tag{4.25}$$

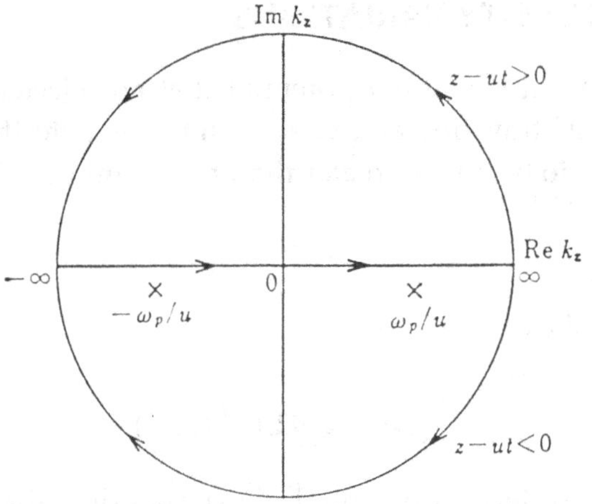

FIGURE 4.4 Contours of k_z integration in Eq. (4.23).

Here, $\theta(x)$ is the unit step function

$$\theta(x)=\begin{cases}1 & (x\le 0)\\ 0 & (x<0)\end{cases} \tag{4.26}$$

and

$$\lambda_p = u/\omega_p \tag{4.27}$$

represents the characteristic length corresponding to the distance ($=2\pi\lambda_p$) over which an ion travels in one period of the plasma oscillation.

Equation (4.25) contains a number of remarkable features. First, we note the involvement of the step function $\theta(ut-z)$: Since ut is the z coordinate of the injected ion, the electron density induced by this ion is limited to the domain $z<ut$ behind it; the effect does not reach the domain $z>ut$. This is a manifestation of the principle of causality that was mentioned in Sec. 2.2.

The induced electron density thus varies with wavelength and frequency characteristic of the plasma oscillation posterior to the injected ion along the z-axis. This is the feature of correlations induced by the ion in the degenerate system of electrons, a feature remarkably different from the case treated in Sec. 1.3.2.

We next take up the issue associated with $\delta(x)\delta(y)$ in (4.25). This function appears because we regarded the injected ion as a point charge and calculated electron-density variations induced by the ion. In the treatment of close encounters between the ion and the electrons, one must regard the particles not as classical point charges but as those having quantum-mechanical spread. In the center-of-mass system for scattering between ion and electron, the reduced mass is that of an electron m; the relative velocity is u because of (4.21a). Hence, the relevant momentum is mu with the corresponding de Broglie wavelength of $\hbar/mu$.

4.3.2 INDUCED POTENTIAL

The space-charge density (4.25) induced in the plasma produces an electrostatic potential associated with it. Taking account of the aforementioned quantum-spread around the z-axis, Vager and Gemmel (1976) calculated the potential as

$$\Phi^{\mathrm{ind}}(R, z-ut) = -\frac{Ze}{\lambda_p}\int_0^{\infty} d\varsigma \frac{\sin(\varsigma/\lambda_p)}{\sqrt{R^2+(\hbar/mu)^2+(\varsigma+z-ut)^2}} \tag{4.28}$$

where $R=\sqrt{x^2+y^2}$ is the distance from the z-axis. As (4.28) represents the electrostatic potential produced by excitation of plasma waves as an ion travels through the plasma, we call it the *wake potential.* Figure 4.5 illustrates such a wake potential computed from (4.28) when a 400 keV proton is injected in a carbon plasma ($\hbar\omega_p = 25.0$ eV).

4.3.3 STOPPING POWER

As we observe in the gradient of the potential at the position of the ion in Figure 4.5, the injected charged particle suffers a retarding force from the wake potential, The *stopping power* of the plasma is then the rate, $-dw/dz$, at which the injected charged particle loses its kinetic energy, w ($=Mu^2/2$), owing to this retardation effect. This rate is thus calculated from (4.28) as

$$-\frac{dw}{dz} = Ze\left.\frac{\partial\Phi^{\mathrm{ind}}}{\partial z}\right|_{\substack{z=ut\\ R=0}} \simeq \frac{Z^2e^2\omega_p^2}{u^2}\ln\left(\frac{1.123mu^2}{\hbar\omega_p}\right). \tag{4.29}$$

Here $\lambda_p \gg \hbar/mu$ is assumed, and $2\exp(-\gamma)=1.123$ where γ is Euler's constant.

Equation (4.29) is the *Bethe formula* for the stopping power. The logarithmic factor contained in it is another example of the Coulomb logarithm mentioned in Sec. 1.3.1.

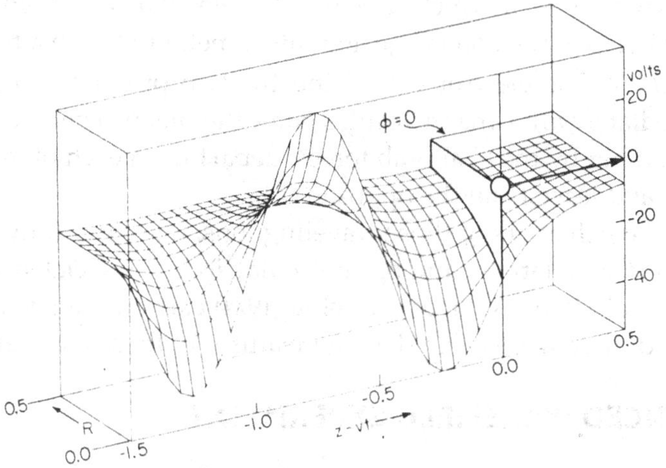

FIGURE 4.5 Wake potential of a 400-keV proton traversing carbon ($\hbar\omega_p = 25.0$ eV). Distances are shown in units of $2\pi\lambda_p = 14.5$ A (Vager & Gemmel, 1976).

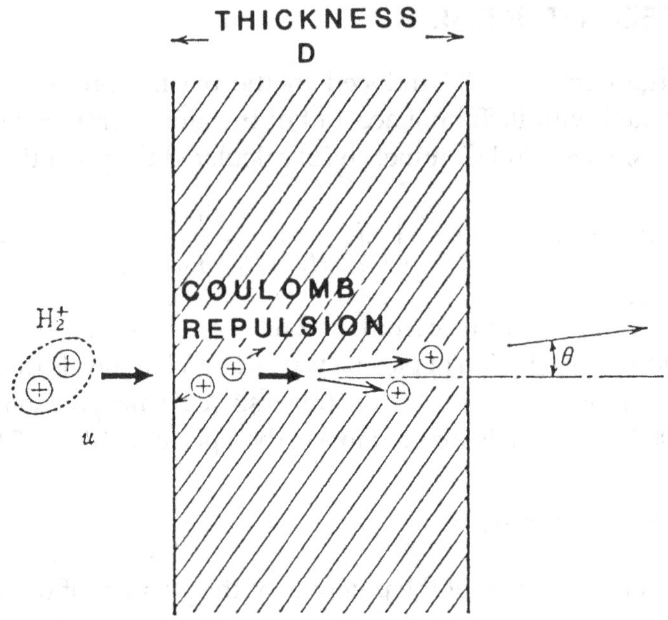

FIGURE 4.6 Ion cluster injected in metal.

4.4 ION CLUSTERS INJECTED IN METALS

Clusters of ions injected in metallic plasmas generate interesting features through the wake potential.

4.4.1 INJECTION INTO THIN FOILS

Consider an experiment in which a beam of molecular ions such as H_2^+ with an energy of ~100 keV per nucleon is injected into a metal foil with a thickness of about 1000 Å (Figure 4.6). The electrons that bind the two protons in H_2^+ molecules are removed immediately after the molecules enter the metal. The two dissociated protons repel each other by the Coulomb forces, depart from each other while traveling through metal, and leave from the opposite side.

Monitoring the behaviors of those traveling ions, one can neatly monitor the features of dynamic correlations among molecular ions, dissociated ions, and metallic electrons. Brandt, Ratkowski, and Ritchie (1974) carried out such stopping-power experiments and thereby revealed detailed features of those correlation effects.

4.4.2 ADVANCED WAKEFIELD EXPERIMENT

The Advanced Wakefield Experiment at CERN, Europe's particle-physics laboratory near Geneva, Switzerland, utilized high-intensity proton clusters—in which each proton had an energy of 400 GeV—to drive a wake potential in a 10-m-long Rb plasma and to accelerate laser-injected electrons by the wake potential up to 2 GeV (Adli et al., 2018).

4.5 X-RAY CRYSTALLOGRAPHY

In the scattering experiments described in Sec. 3.2, if we pay no attention to the frequencies, that is, if we integrate the scattered waves over the frequencies, then the differential cross-section of scattering into a solid angle do may be expressed as

$$\frac{dQ}{do} = \frac{3N}{8\pi}\sigma_T\left(1 - \frac{1}{2}\sin^2\theta\right)S(\mathbf{k}). \tag{4.30}$$

Here, the static structure factor $S(\mathbf{k})$ defined as

$$S(\mathbf{k}) = \frac{1}{N}\left\langle\left|\rho_{\mathbf{k}}(t)\right|^2\right\rangle = \frac{1}{n}\int_{-\infty}^{\infty} d\omega S(\mathbf{k},\omega) \tag{4.31}$$

corresponds to the spectral distribution of spatial density fluctuations (e.g., Ichimaru, 1973), that describes spatial density configurations such as the lattice structures.

Short-ranged crystalline order at nearest-neighbor separations may thus be approached by these scattering techniques (Figure 4.1) through (4.30).

In fact, von Laue observed such diffraction patterns in 1914 by shining X-ray onto metal; the father-and-son Braggs then developed *X-ray crystallography* in 1915. Both works led to the Nobel Prizes in the respective years.

Recently, the advent of accurate X-ray scattering techniques has made it possible to measure the physical properties of dense plasmas for study in high-energy density physics (Glenzer & Redmer, 2009), and we shall revisit these subjects subsequently in Sec. 4.7.

4.6 OBSERVATION OF LAUE PATTERNS IN COULOMB GLASSES

In conjunction with the aforementioned scattering experiments, we now turn to the observation of layered structures and Laue patterns in Coulomb glasses, created by the Monte Carlo (MC) simulations (Ogata & Ichimaru, 1989) of Sec. 2.4.1. In fact, solidifications such as crystallization and/or glass transition are intriguing events in the thermal evolution of many-particle systems.

Later, in Sec. 5.2, we also investigate conditions that a one-component plasma (OCP) fluid crystallizes into a Coulomb solid through comparison of the free energies in the respective phases, as temperature is lowered below $\Gamma > 180$.

4.6.1 MADELUNG ENERGY

When an OCP forms a crystalline lattice with a lattice constant b, the electrostatic energy per particle E_M, called the *Madelung energy*, is expressed as

$$E_M = -\frac{1}{2}\alpha_M \frac{(Ze)^2}{b},$$

where α_M is called the *Madelung constant*. Table 4.1 lists the values of the energy constant for various lattices (Foldy, 1978).

It has been assumed that an OCP in its ground state forms a body-centered-cubic (bcc) crystalline solid, a conclusion reached through comparison of the Madelung energies of the several cubic-lattices and other structures. Extensive MC simulations have been performed for OCP solids with cubic structures (Brush, Sahlin, & Teller, 1966; Slattery, Doolen, & DeWitt, 1982), and the bcc lattice has been shown to have the lowest free energy at finite temperatures as well.

As we find in Table 4.1, however, the differences between the energy constants for those cubic lattices are so small that it would be inconceivable to assume a monocrystalline structure formed in a real solid.

In fact, if a rapid quench is applied to an OCP fluid from a temperature in excess of $\Gamma = 180$, the resultant state might possibly be a Coulomb glass, characterized by random polycrystalline structures with long-ranged bond-orientational order; such a conjecture stems from the fact that the differences in energies between various lattices are so minute. The particles may then be viewed as virtually locked around their positions in quasi-equilibria.

4.6.2 LAYERED STRUCTURES AT VARIOUS QUENCHES

We thus follow the dynamic evolution of an OCP by MC simulations with 432 particles, starting with a fluid state at $\Gamma = 160$, leading to the formation of Coulomb glasses at different quenches: (A) an application of a sudden quench to $\Gamma = 400$ at $c = 0$; (B) an application of a gradual quench stepwise with $\Delta\Gamma = 10$ from $\Gamma = 160$ at $c = 0$ to $\Gamma = 400$ at $c = 23$; (C) a sudden quench to $\Gamma = 300$ at $c = 0$ (Ogata & Ichimaru, 1989). Here, c denotes the sequential number of MC configurations measured in units of a million configurations; the sequential number corresponds to an MC time via $\omega_p t = 2.7 \times 10^2 c$, with ω_p referring to the plasma frequency (1.27).

To study the nature of interlayer correlations, we identify the particles in the three central layers and project their positions normally onto a plane. In Figure 4.7 on the left, such projections are exhibited for the quenches (A)–(C), where open circles, closed circles, and crosses denote projections of the particles on upper, middle, and lower layer, respectively.

TABLE 4.1 Energy Constants for Various Lattices

Lattice	$\alpha_M a/b$
Body-centered-cubic (bcc)	−1.791 858 52
Face-centered-cubic (fcc)	−1.791 747 23
Hexagonal-close-packed (hcp)	−1.791 676 90
Simple cubic (sc)	−1.760 118 90
Diamond (dmnd)	−1.670 851 41

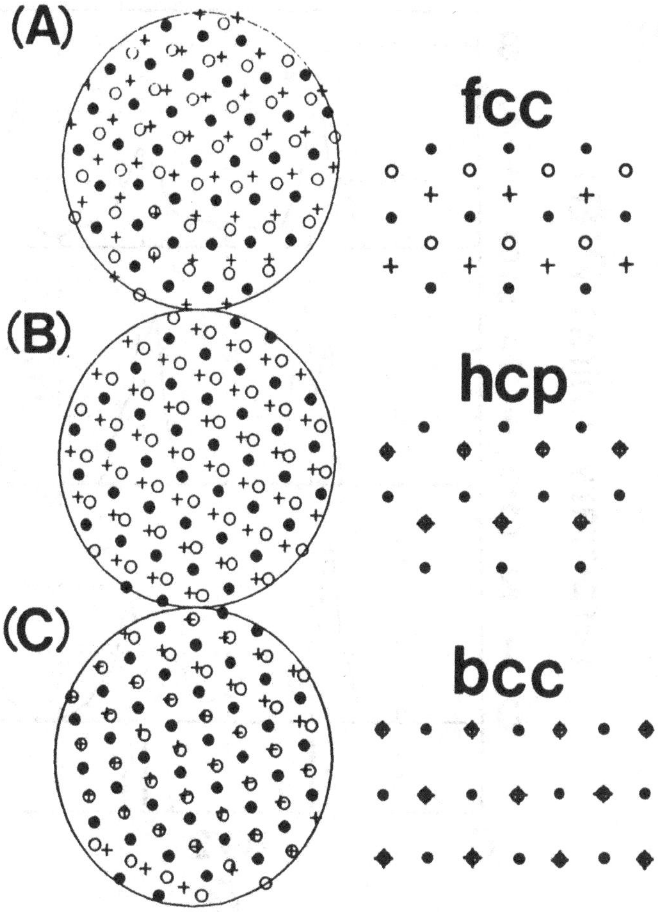

FIGURE 4.7 Normal projections of most closely packed layers: *Open circles,* upper layer; *closed circles,* middle layer; *crosses,* lower layer (Ogata & Ichimaru, 1989).

For comparison, analogous projections are shown in Figure 4.7 on the right for the most closely packed layers in the face-centered-cubic (fcc), hexagonal-close-packed (hcp) and bcc crystals. We find here that the Coulomb glass with the quench (B) has developed an advanced state of polycrystalline nucleation predominantly with local fcc-hcp configurations over those with (A) and (C).

Intralayer correlations are investigated in terms of the bond-angle distributions $P(\theta)$ (Figure 4.8) and the two-dimensional radial distribution functions $g(r)$ (Figure 4.9); these are joint probability densities of finding two particles at a separation r. We define "bonds" as those lines connecting two adjacent particles located within $r < 2.5$ on a layer, "bond angle" as the angle between a pair of such bonds originating from a particle, and "coordination number" (CN) as the total number of bonds originating from a given particle.

In Figure 4.8, we plot the bond-angle distribution functions between intralayer particles in the quenched states (A)–(C) and compare them with analogous quantities for the fcc-hcp (i.e., hexagonal) and bcc lattices. In the state (B), $P(\theta) = 0$ observed at $\theta \sim \pi/2$

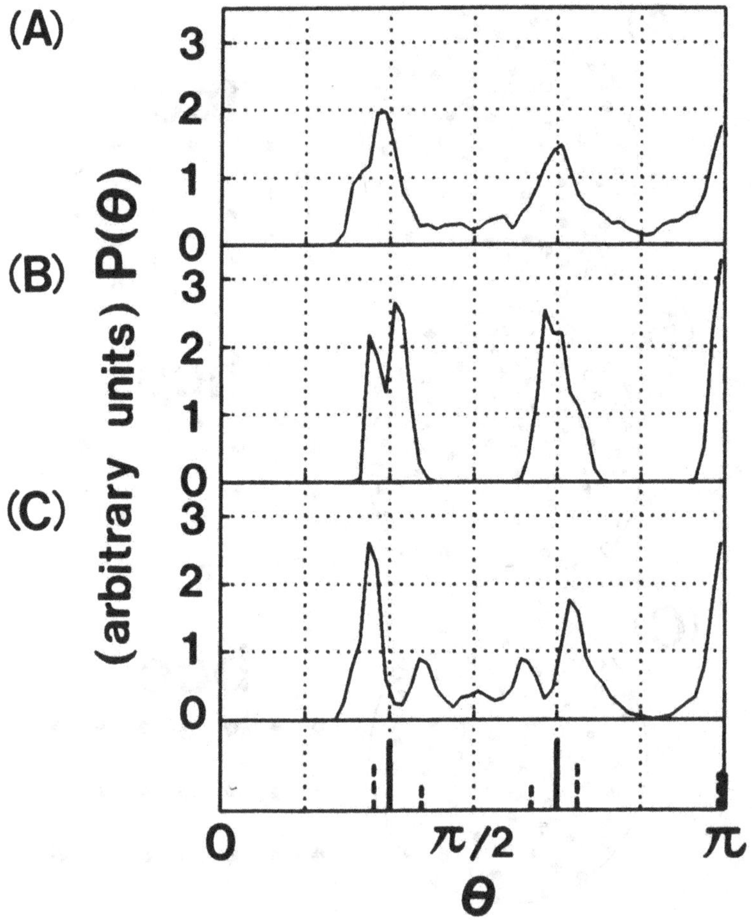

FIGURE 4.8 Bond-angle distribution functions between intralayer particles in the glasses (A)–(C). In *the bottom* of the figure, the *solid lines* indicate the bond angles for the fcc-hcp (hexagonal) lattices; the *dashed lines*, those for the bcc lattices (Ogata & Ichimaru, 1989).

and $5\pi/6$ implies an advanced degree of nucleation, while splitting of the peaks at $\theta \sim \pi/3$ and $2\pi/3$ indicates disordering effects on the local hexagonal configurations.

In Figure 4.9, we plot the two-dimensional radial distribution functions between intralayer particles and compare them with the peak positions for the fcc-hcp and bcc lattices. As seen in the figure, all the particles in the state (B) have CN = 6, implying little distortion in the local hexagonal configurations. In the states (A) and (C), however, distortion in the hexagonal configurations is substantial, since 89% and 90%, respectively, of the particles have CN = 6, while 5% and 8% have CN = 5, and 6% and 2% have CN = 7.

4.6.3 LAUE PATTERNS FOR GLASSES

Finally, we investigate the combined effects between the intralayer and interlayer correlations by a scattering method of Sec. 4.5. We thus inject plane waves with wave vector $\mathbf{k}_1$ to the glasses (A)–(C) in the direction normal to the layered structures

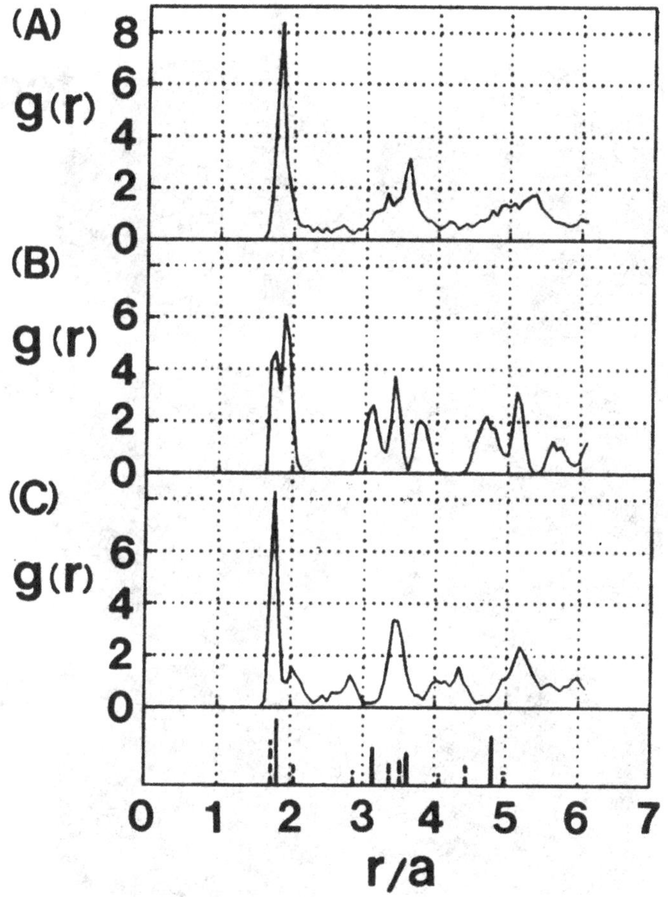

FIGURE 4.9 Two-dimensional radial distribution functions between intralayer particles in the glasses (A)–(C). The *bottom* of the figure shows the peak positions for the fcc-hcp (*solid lines*) and bcc (*dashed lines*) lattices (Ogata & Ichimaru, 1989).

and measure the strength of scattered waves $\mathbf{k}_2$ in the directions specified by (χ, ϕ), where χ is the scattering angle between $\mathbf{k}_1$ and $\mathbf{k}_2$ and ϕ is the azimuthal angle around $\mathbf{k}_1$.

The cross-section (4.30) for coherent scattering is proportional to the static structure factor (4.31), where $\mathbf{k}=\mathbf{k}_2-\mathbf{k}_1$. We assume that the incident wave numbers have a distribution proportional to $\exp[-(k_1-k_0)^2/\kappa^2]$ with $k_0=2\pi$ and $\kappa=0.24$. The scattering experiment is thus capable of detecting the coherence in the phases $4\pi\sin(\chi/2)\mathbf{k}\cdot\mathbf{r}_j/k$ over those particles $\mathbf{r}_j$ contained in a slab of width $4.2/\sin(\chi/2)$ in the direction of $\mathbf{k}$.

Figure 4.10 displays the Laue patterns obtained for the glass states (A)–(C), and compares them with those of the fcc, hcp, and bcc lattices. We observe the existence of local hexagonal order in (B) and to a lesser extent in (A); a slight involvement of local bcc configurations is likewise detected for all the cases of (A)–(C).

In light of the analyses described above, we may conjecture the following stages of evolution for the glass transitions in dense plasmas: In a super-cooled OCP without

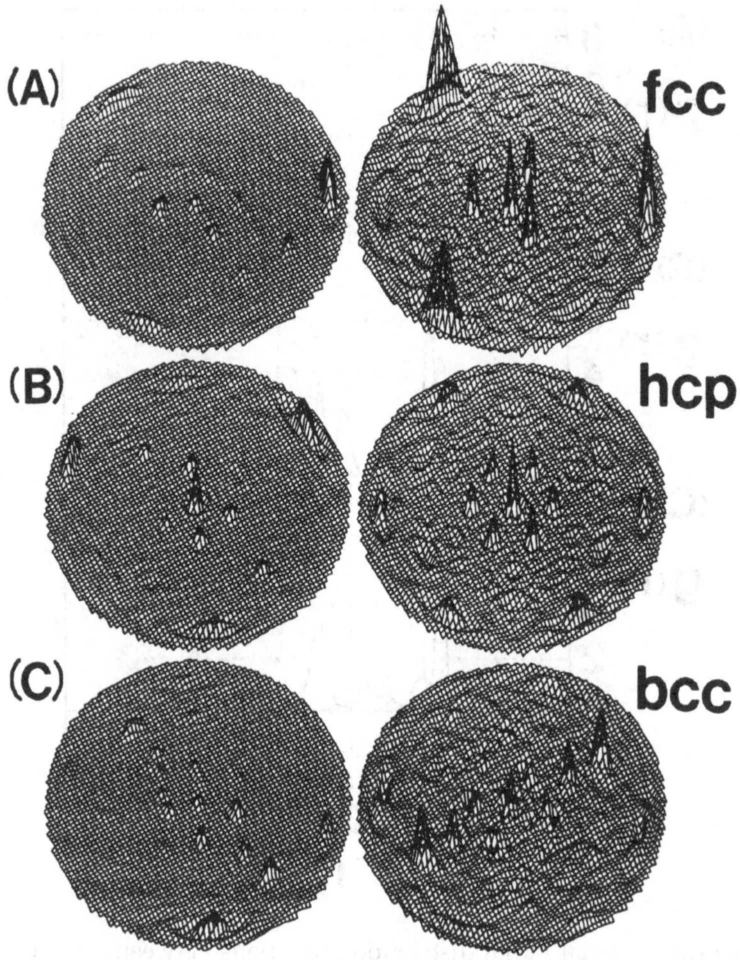

FIGURE 4.10 Laue patterns for the glasses (A)–(C) and for the fcc, hcp, and bcc lattices of 432 particles. The polar coordinates consist of $0 \leq (\pi - \chi)/2 \leq 0.45\pi$ and $0 \leq \phi \leq 2\pi$; the origin corresponds to $\chi = \pi$. Here, χ is the scattering angle between incident and scattered waves; ϕ is the azimuthal angle around the incident wave (Ogata & Ichimaru, 1989).

an external field, layered structures (i.e., a one-dimensional order) develop first in an arbitrary direction. Intralayer (i.e., two-dimensional) ordering then follows, which would favor formation of fcc-hcp (i.e., hexagonal) local clusters.

Since the bcc lattice has a Madelung energy slightly lower than the fcc or hcp lattice in Coulombic systems (Table 4.1), a possibility of nucleation remains for bcc clusters. Hence, the resultant state may have a complex polycrystalline structure.

Later, in Sec. 10.3, we shall consider first-principles calculations of shear moduli for Monte-Carlo-simulated Coulomb solids, with the inclusion of the Coulomb glasses, and apply the results for improved analyses of the non-radial oscillations in neutron stars.

4.7 X-RAY THOMSON SCATTERING AND TIME-RESOLVED XANES DIAGNOSTIC WITH HIGH ENERGY DENSITY PLASMAS

Accurate measurements of the states of plasmas, including temperature, density, and ionization in dense matter, are essential in high-energy density physics (Glenzer & Redmer, 2009).

In Sec. 3.6, we remarked on the X-ray Thomson scattering measurements in the collective regime carried out in a beryllium target with solid density that was heated isochorically with a broadband X-ray source into a state of dense plasma with weakly degenerate electrons; the characteristic peak associated with the collective plasma oscillation was thereby observed. The spectrally resolving X-ray scattering technique has also been applied in a number of laboratories to study the properties of such dense plasmas (Sawada et al., 2007; Ravasio et al., 2007).

Solid-to-plasma transition dynamics have been approached with the aid of a recently advanced diagnostic technique such as time-resolved X-ray near edge spectroscopy (XANES) (Dorchies & Recoules, 2016). Electronic and structural properties with three different (simple, transition, and noble) types of metals have been investigated through absorption spectroscopy experiments with the aid of ultrafast X-ray free electron lasers (Dorchies et al., 2008; Cho et al., 2011; Katayama et al., 2013; Gaudin et al., 2014).

5

THERMODYNAMICS OF CLASSICAL OCP AND QUANTUM ELECTRON LIQUIDS

The microscopic descriptions of the plasma in terms of the correlation functions and the structure factors are connected directly to the thermodynamics that specifies the macroscopic states and/or the phases of the system. Associated with these is the emergence of features such as insulator-to-metal transition, order–disorder transition, para- to ferromagnetic transition, and chemical separation. To lay foundations for such phase analyses, we now study the thermodynamic functions for the classical one-component plasma (OCP) as well as for the quantum liquids of electrons.

5.1 RADIAL DISTRIBUTION FUNCTIONS AND CORRELATION ENERGIES

The radial distribution function $g(\mathbf{r})$ is a joint probability density of finding two particles at a separation r. As was described in Sec. 2.1.2, it is related directly to the static structure factor $S(\mathbf{k})$ in (2.8) as

$$g(\mathbf{r}) = 1 + \frac{1}{n}\int \frac{d\mathbf{k}}{(2\pi)^3}(S(\mathbf{k}) - 1)\exp(i\mathbf{k}\cdot\mathbf{r}). \tag{5.1}$$

Figure 5.1 exhibits g(**r**) evaluated for $\Gamma > 1$ with the aid of the Monte Carlo (MC) simulation methods explained in Sec. 2.4.1 (Iyetomi, Ogata, & Ichimaru 1992). Analytically, one can also evaluate the radial distribution functions through the solution to a set of integral equations. Figure 5.1 shows the results of such an integral-equation scheme,

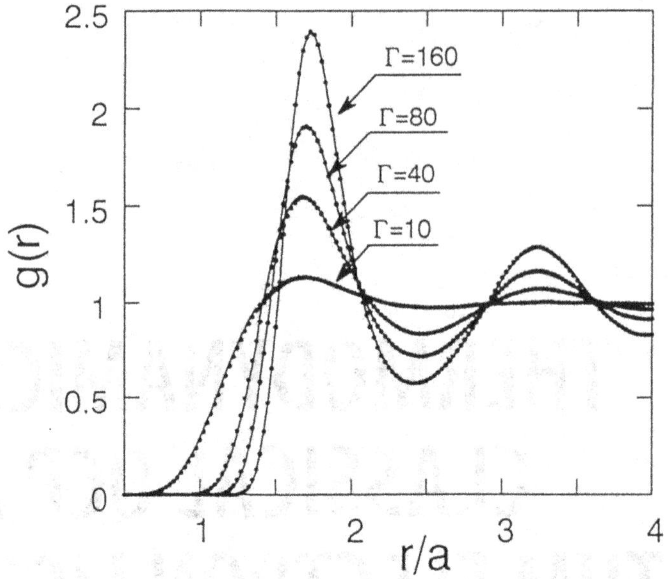

FIGURE 5.1 Radial distribution functions of OCP fluids obtained by MC methods with $N=1024$ at various values of Γ. The number of the MC configurations generated for each run was 7×10^6; $g(r)$ was calculated with 200 bins in the range $0\leq r\leq L/2$, a half of the cubic MC cell with size $L=16.2a$. The solid curves represent the results calculated with the improved hypernetted-chain scheme (Iyetomi, Ogata, & Ichimaru, 1992).

called the improved hypernetted-chain (e.g., Ichimaru, Iyetomi, & Tanaka, 1987) calculations; agreements appear excellent.

The correlation energy U_{int} per unit volume can then be calculated, once either $S(\mathbf{k})$ or $g(\mathbf{r})$ is known, through formulae,

$$\begin{aligned} U_{\text{int}} &= 2\pi(Ze)^2 n \int \frac{d\mathbf{k}}{(2\pi)^3}\frac{1}{k^2}[S(\mathbf{k})-1] \\ &= \frac{(Zen)^2}{2}\int d\mathbf{r}\frac{1}{r}[g(\mathbf{r})-1]. \end{aligned} \tag{5.2}$$

Thermodynamics of the plasma may be approached through evaluation of such quantities.

5.1.1 CORRELATION ENERGY IN THE RPA

Substituting the random-phase approximation (RPA) structure factor (3.20b) in (5.2), we obtain the RPA expression for the normalized correlation energy, $u_{\text{ex}} \equiv U_{\text{int}}/nk_BT$, as

$$u_{\text{ex}}^{\text{DH}} = -\frac{\sqrt{3}}{2}\Gamma^{3/2}. \tag{5.3}$$

This is the RPA correlation energy, called the Debye–Hückel contribution.

5.1.2 MULTI-PARTICLE CORRELATION

The RPA correlation energy (5.3) takes account of binary correlation to the lowest order in Γ and thus is applicable to OCP only for $\Gamma \ll 1$. The expression for the correlation energy next order in the Γ expansion has also been determined precisely as

$$u_{\mathrm{ex}}^{\mathrm{ABE}} = -\frac{\sqrt{3}}{2}\Gamma^{3/2} - 3\Gamma^{3}\left[\frac{3}{8}\ln(3\Gamma) + \frac{\gamma}{2} - \frac{1}{3}\right] \quad (\Gamma < 0.1) \tag{5.4}$$

with $\gamma = 0.57721\ldots$ denoting Euler's constant.

Correlation energies beyond RPA require accurate assessment of the triple- and higher-order correlations. These have been approached through various theoretical methods, including the giant cluster-expansion calculation (Abe, 1959), expansion in Γ of the Bogoliubov-Born-Green-Kirkwood-Yvon (BBGKY) hierarchy (O'Neil & Rostoker, 1963), multi-particle correlation in the convolution approximation (Totsuji and Ichimaru, 1973), and the improved hypernetted-chain integral-equation scheme based on the density-functional formulation of multi-particle correlations (Ichimaru, Iyetomi, & Tanaka, 1987; Ogata & Ichimaru, 1987; Iyetomi, Ogata, & Ichimaru, 1992).

5.2 OCP THERMODYNAMIC FUNCTIONS

Earlier, in Sec. 2.2.2, we introduced thermodynamic functions for the study of phase transitions and related phenomena in plasmas.

The correlation energies in large Γ regime (Slattery, Doolen, & DeWitt, 1982; Ogata & Ichimaru, 1987) may be evaluated by the use of Monte Carlo radial distribution functions, such as those in Figure 5.1, to yield:

$$u_{\mathrm{ex}}^{\mathrm{OCP}} = -0.898004\Gamma + 0.96786\Gamma^{1/4} + 0.220703\Gamma^{-1/4} - 0.86097 \quad (1 < \Gamma < 180). \tag{5.5}$$

In the intermediate-coupling regime, $0.1 \le \Gamma < 1$, the excess internal energy has been calculated (Slattery, Doolen, & DeWitt, 1980) through the solution to the hypernetted-chain (e.g., Ichimartu, Iyetomi, & Tanaka, 1987) scheme.

With the aid of these calculations, one finds a formula connecting (5.4) and (5.5) as (Ichimaru, 2004b)

$$u_{\mathrm{ex}}(\Gamma) = \frac{u_{\mathrm{ex}}^{\mathrm{ABE}}(\Gamma) + 3\times 10^{3}\Gamma^{5.7}u_{\mathrm{ex}}^{\mathrm{OCP}}(\Gamma)}{1 + 3\times 10^{3}\Gamma^{5.7}} \quad (\Gamma < 180). \tag{5.6}$$

This formula is applicable for a classical OCP fluid in the range $\Gamma < 180$ with the accuracy better than 0.1%.

5.2.1 OCP FREE ENERGY

The free energy per particle, $f(\Gamma)$, in units of the thermal energy is then given as a sum of the ideal-gas and excess contributions (Ichimaru, 2004b):

$$f(\Gamma) \equiv \frac{F}{nk_BT} = 3\ln(\Lambda) - 1 + \ln\left(\frac{3}{4\pi}\right) + f_{\text{ex}}(\Gamma) \tag{5.7}$$

with

$$\Lambda = \frac{\hbar}{a}\left(\frac{2\pi}{mk_BT}\right)^{1/2}. \tag{5.8}$$

representing the ratio between the thermal de Broglie wavelength and the ion-sphere radius. In (5.7), the excess free energy, $f_{\text{ex}}(\Gamma)$, is calculated through the coupling-constant integration (2.30) of (5.6) as

$$\begin{aligned} f_{\text{ex}}^{\text{OCP}}(\Gamma) = &-0.898004\Gamma + 3.87144\Gamma^{1/4} - 0.882812\Gamma^{-1/4} \\ &-0.86097\ln\Gamma - 2.52692. \end{aligned} \tag{5.9}$$

In (2.30), $U_{\text{int}}(\eta)$ refers to the correlation energy (5.6) evaluated in a system where the strength of Coulomb coupling Γ is replaced by $\eta\Gamma$.

5.2.2 OCP PRESSURE

The excess pressure P_{ex} may be evaluated by differentiation of F_{ex} with respect to volume V at a constant temperature, that is,

$$P_{\text{ex}} = -\left(\frac{\partial F_{\text{ex}}}{\partial V}\right)_{T,N}. \tag{5.10}$$

5.2.3 SOLID FREE ENERGY

The thermodynamic functions for classical OCP solids have been investigated through the MC simulations coupled with analytic study of the anharmonic effects in the lattice vibrations (Dubin, 1990). The normalized correlation energy for an OCP body-centered-cubic (bcc)-crystalline solid is thus given by

$$u_{\text{ex}}^{\text{OCP}}(\Gamma) = -0.895929\Gamma + 1.5 + \frac{10.84}{\Gamma} + \frac{352.8}{\Gamma^2} + \frac{1.74\times10^5}{\Gamma^3},$$

where the first term on the right-hand side reflects the value of the Madelung energy for the bcc lattice (Foldy, 1978); it is the electrostatic energy per particle in an OCP that forms a crystalline lattice.

The free energy $f(\Gamma)$ in the classical OCP solid can be evaluated by integrating the correlation energy with respect to the inverse temperature as (Dubin, 1990)

$$f^{\text{OCP}}(\Gamma) = -0.895929\Gamma + \frac{9}{2}\ln\Gamma - 1.885596 - 10.84\Gamma^{-1} - 176.4\Gamma^{-2} - 5.980\times10^{4}\Gamma^{-3} \tag{5.11}$$

under the assumption that the ground-state free energy $f(\infty)$ is given by the harmonic lattice value (Pollock & Hansen, 1973).

5.2.4 WIGNER CRYSTALLIZATION

Comparing the free energies between the fluid and crystalline phases, we find that an OCP fluid may freeze (i.e., Wigner transition) into a bcc crystal at $\Gamma = 172 \sim 180$. In light of a possible formation of Coulomb glasses, as considered in Sec. 4.6, however, how the actual transitions may take place in real plasmas remains a delicate issue.

5.3 EQUATIONS OF STATE FOR QUANTUM ELECTRON LIQUIDS

Metallic hydrogen is a binary system of itinerant electrons and those protons in a fluid or in a solid state. In the jellium model of metals, we regard those itinerant electrons as a quantum electron liquid (Pines & Nozièrez, 1966). Basic parameters for such electron liquids at metallic densities have been defined in Sec. 1.2.

5.3.1 IDEAL-GAS CONTRIBUTIONS

The Helmholtz free energy and the pressure are expressed as sums of the ideal-gas and exchange-correlation parts:

$$f(\Gamma,\theta) = f_0(\theta) + f_{\text{xc}}(\Gamma,\theta). \tag{5.12a}$$

$$p(\Gamma,\theta) = p_0(\theta) + p_{\text{xc}}(\Gamma,\theta). \tag{5.12b}$$

The Gibbs free energy is then given by

$$G = F + PV. \tag{5.12c}$$

The ideal-gas contribution to the free energy is expressed as a balance between those of the chemical potential, $\mu(\Gamma,\theta)$ (= Gibbs free energy per particle) and the pressure as

$$f_0(\theta) = \mu_0(\theta) - p_0(\theta). \tag{5.13}$$

Through numerical investigation of the relevant Fermi integrals (cf. Appendix IV), we find that the chemical potential may be accurately fitted by the expression,

$$\mu_0(\theta)=\ln\left(\frac{4}{3\sqrt{\pi}}\theta^{-3/2}\frac{1+1.01820\theta^{-3/2}}{1+0.75225\theta^{-3/2}+0.76595\theta^{-3}}\right)$$
$$+\frac{1}{\theta+0.82247\theta^{3}+0.23645\theta^{3.5}}. \tag{5.14}$$

Similarly, we obtain for the pressure,

$$p_0(\theta)=1+\frac{2}{5\theta}\frac{1-1.78600\theta+2.32734\theta^2}{1+0.71400\theta+2.14768\theta^{2.5}}. \tag{5.15}$$

Both of the preceding expressions exactly satisfy the limiting conditions for the first two terms in the expansions for $\theta \ll 1$, i.e., $\mu_0=E_F$ and $p_0=(2/5)nE_F$, as well as for $\theta \gg 1$, i.e.,

$$\frac{\mu_0}{k_BT}=-\frac{3}{2}\ln\theta+\ln\frac{4}{3\sqrt{\pi}}$$

and $p_0=nk_BT$.

5.3.2 EXCHANGE–CORRELATION CONTRIBUTIONS

The exchange–correlation energies of electron liquids at finite temperatures $\theta=0.1, 1.0, 5.0$ were evaluated through a solution to a set of integral equations (Tanaka & Ichimaru, 1986). The results were then parameterized in analytic formulae as (Ichimaru, Iyetomi, & Tanaka, 1987; Ichimaru, 2004b)

$$u_{xc}(\Gamma,\theta)=-\Gamma\frac{a(\theta)+b(\theta)\sqrt{\Gamma}+c(\theta)\Gamma}{1+d(\theta)\sqrt{\Gamma}+e(\theta)\Gamma}. \tag{5.16}$$

with

$$a(\theta)=a^{\mathrm{HF}}(\theta)=\left(\frac{3}{2\pi}\right)^{2/3}\frac{0.75+3.04363\theta^2-0.09227\theta^3+1.7053\theta^4}{1+8.31051\theta^2+5.1105\theta^4}\tanh\left(\frac{1}{\theta}\right), \tag{5.17a}$$

$$b(\theta)=\frac{0.341308+12.070873\theta^2+1.148889\theta^4}{1+10.495346\theta^2+1.326623\theta^4}\sqrt{\theta}\tanh\left(\frac{1}{\sqrt{\theta}}\right), \tag{5.17b}$$

$$c(\theta)=\left[0.872756+0.025248\exp\left(-\frac{1}{\theta}\right)\right]e(\theta), \tag{5.17c}$$

$$d(\theta) = \frac{0.614925 + 16.996055\theta^2 + 1.489056\theta^4}{1 + 10.10935\theta^2 + 1.22184\theta^4}\sqrt{\theta}\tanh\left(\frac{1}{\sqrt{\theta}}\right), \tag{5.17d}$$

$$e(\theta) = \frac{0.539409 + 2.522206\theta^2 + 0.178484\theta^4}{1 + 2.555501\theta^2 + 0.146319\theta^4}\theta\tanh\left(\frac{1}{\theta}\right). \tag{5.17e}$$

The term (5.17a) represents the Hartree–Fock contribution (Perrot & Dharma-wardana, 1984).

The coupling-constant integration (2.30) is performed with (5.16) to yield

$$\begin{aligned} f_{\mathrm{xc}}(\Gamma,\theta) = & -\frac{c}{e}\Gamma - \frac{2}{e}\left(b - \frac{cd}{e}\right)\sqrt{\Gamma} - \frac{1}{e}\left[\left(a - \frac{c}{e}\right) - \frac{d}{e}\left(b - \frac{cd}{e}\right)\right]\ln\left|e\Gamma + d\sqrt{\Gamma} + 1\right| \\ & + \frac{2}{e\sqrt{4e - d^2}}\left[d\left(a - \frac{c}{e}\right) - \left(2 - \frac{d^2}{e}\right)\left(b - \frac{cd}{e}\right)\right] \\ & \times\left[\tan^{-1}\left(\frac{2e\sqrt{\Gamma} + d}{\sqrt{4e - d^2}}\right) - \tan^{-1}\left(\frac{d}{\sqrt{4e - d^2}}\right)\right]. \end{aligned} \tag{5.18}$$

A condition that $4e - d^2 > 0$ is satisfied for any θ. Values of the exchange and correlation free energy in the ground state, given by (5.18), agree accurately with those obtained by Green's function Monte Carlo simulations (Ceperley & Alder, 1980) at $r_s = 2, 5, 10, 20, 50, 100$.

5.3.3 ORIGIN OF COHESIVE FORCES

It is instructive to examine parameter dependence and sign of the elementary contributions in the specific pressure (5.12b) in the limit of the quantum degeneracy, $\theta \to 0$. The ideal-gas contributions behave as

$$p_0 \to 2/5\theta \sim n^{2/3} \quad (> 0). \tag{5.19a}$$

The Hartree–Fock contribution to the pressure (5.10) stems from the evaluation (5.16) in which only the term (5.17a) is retained, and it takes on the value,

$$p_{\mathrm{xc}}^{\mathrm{HF}} \to -0.083 r_s/\theta \sim e^2 n^{1/3} \quad (< 0). \tag{5.19b}$$

The Coulomb pressure, which represents the large-r_s contributions in (5.10), likewise behaves as

$$p_{\mathrm{xc}}^{\mathrm{Coul}} \to -0.158 r_s/\theta \sim e^2 n^{1/3} \quad (< 0). \tag{5.19c}$$

Note, both the exchange and Coulomb terms, (5.19b) and (5.19c), are negative and proportional to the strength of Coulomb coupling represented by e^2.

We specifically emphasize the significance of these observations in elucidating the origin of cohesive forces in the ferromagnetic and freezing transitions.

5.4 FREEZING AND FERROMAGNETIC TRANSITIONS IN ELECTRON LIQUID

Thus far, we have considered itinerant electrons forming an electron liquid and protons in a fluid or solid state. Metallic hydrogen, which we consider subsequently, is a binary system of those electrons and protons.

The electron liquid is a quantum OCP of electrons immersed in a uniform compensating background of positive charges. Electrons are fermions, with spin 1/2, obeying the Fermi statistics. Wave functions of two identical fermions with parallel spins are antisymmetric, that is to say, they change their signs when the positions of the two fermions are interchanged. The values of wave functions vanish when two identical fermions occupy the same position; interpreted physically, identical fermions with parallel spins repel each other.

This observation then accounts for the origin of the spin-discriminating (repulsive) *exchange forces* between such identical fermions. These exchange forces and the ordinary Coulomb forces, both repulsive, are effective between protons as well as between electrons.

These repulsive forces induce the so-called "exchange" and "Coulomb" holes in the two-particle distribution functions for electrons (e.g., Ichimaru, 1982). The interaction between electrons and such "holes," which is attractive, then produces negative contributions to the free energy and the pressure, as (5.19b) and (5.19c) exemplify. These negative free energies stimulate spin ordering in a ferromagnetic transition and crystalline ordering in a freezing transition.

It has thus been expected that an electron liquid may undergo a *magnetic transition,* from a spin-non-aligned, paramagnetic phase to a spin-aligned, ferromagnetic phase (Ceperley & Alder, 1980; Ichimaru, 1997, 2000; Ortiz, Harris, & Ballone, 1999) near the conditions for Wigner crystallization, a phase transition of dilute electrons into a crystalline state at low temperatures (Wigner, 1935, 1938). A magnetic transition takes place basically through competition between the spin-dependent exchange processes, which favor a ferromagnetic state, and the kinetic energies, which favor a paramagnetic state.

Analogous situations may exist in the case of a *freezing transition,* where the repulsive Coulomb forces favor an inhomogeneous distribution such as one in a Wigner crystal, while the kinetic processes favor a uniform distribution characteristic of a fluid state.

6

PHASE DIAGRAMS OF HYDROGEN

Hydrogen, which we know as a light gaseous substance at ambient temperature and pressure, may exhibit extraordinary features pertinent to the strongly correlated plasmas when it is compressed to densities comparable to or greater than those of ordinary solids. Basically, hydrogen matter is a statistical ensemble consisting of electrons and protons. The protons, with the smallest atomic number (unity) among various chemical elements and thus with de Broglie wavelengths longer than those of other nuclear species, tend to interfere more conspicuously with each other quantum mechanically under such condensed circumstances.

Dense hydrogen under ultrahigh pressures as found in stellar and planetary interiors may be expected to undergo transformation between phases such as *metallization, crystallization,* and *magnetization.* All of these phase transitions are not only of great interest in the condensed-matter physics, but, since hydrogen is the most abundant chemical element in the Universe, their nature crucially affects fundamental issues in astrophysics, such as the generation of energy and magnetism in the interiors of stars and planets as well as energy transport to stellar and planetary surfaces. Thus, the physics of hydrogen constitutes a vital element in the formation, structure, and evolution of these astronomical objects.

6.1 STATES OF HYDROGEN

A hydrogen atom is a bound state between an electron and a proton. The orbital radius of a bound electron in the ground state is the *Bohr radius,* given by

$$a_B = \frac{\hbar^2}{me^2} = 0.529177 \text{ Å}. \tag{6.1a}$$

The binding energy of a hydrogen atom in the ground state constitutes a unit of energy called the *Rydberg* and takes on the value,

$$\frac{me^4}{2\hbar^2} = 13.6057 \text{ eV} \equiv 1 \text{ Ry}. \tag{6.1b}$$

These provide typical scales of length and energy in the atomic physics of hydrogen.

6.1.1 MOLECULAR HYDROGEN

A hydrogen molecule is a bound state between two hydrogen atoms; in the ground state, the average interproton spacing is 0.742 Å ($\approx 1.4\, a_B$).

The dissociation energy and the ionization potential of a hydrogen molecule are 4.474 eV (≈0.33 Ry) and 15.43 eV (≈1.13 Ry), respectively. The dissociation energy of a molecular ion, H_2^+, is 2.467 eV (≈0.18 Ry).

6.1.2 PRESSURE IONIZATION

If the number density n of protons is high so that the Wigner–Seitz radius (1.6) is less than the Bohr radius, i.e., $a < a_B$ (corresponding to $n > 1.6 \times 10^{24}$ cm^{-3}), wave functions of orbital electrons in neighboring hydrogen atoms or molecules significantly overlap each other, and so they make conduction electrons; such a process is called *pressure ionization.*

In terms of the Fermi energy E_F and the Fermi wave number k_F,

$$E_F = (\hbar k_F)^2/2m = \hbar^2 (3\pi^2 n_e)^{2/3}/2m, \tag{6.2a}$$

of the electrons with number density n_e, the Fermi pressure is calculated as

$$P_F = \frac{2}{5} n_e E_F \approx 51.15 \left(\frac{n_e}{1.6 \times 10^{24} \text{ cm}^{-3}} \right)^{5/3} \text{ (Mbar)}, \tag{6.2b}$$

implying a pressure in a multi-megabar range for the pressure ionization.

In 1935, E. Wigner and H. B. Huntington were the first to predict the possibility of such a metallic modification of hydrogen at an extreme pressure. They did so through calculations of the energy of a body-centered lattice of hydrogen as a function of the lattice constant and by comparison of the result with the energy of the molecular form (Wigner & Huntington, 1935).

Hydrogen is thus expected to undergo a first-order, metal–insulator (MI) transition at an ultrahigh density or in a pressure range of megabars (Ceperley & Alder, 1987; Kitamura & Ichimaru, 1998; McMahon et al., 2012). We shall revisit these subjects in subsequent sections.

6.1.3 LABORATORY REALIZATION OF METALLIC HYDROGEN

Ultrahigh-pressure metal physics experiments have been undertaken for laboratory realization of such metallic hydrogen and for the elucidation of the equations of state and the transport properties of dense hydrogen.

The experimental approaches include diamond-anvil-cell compression (e.g., Mao & Hemley, 1989, 1994) and shock compression (e.g., Dick & Kerley, 1980; Mitchell & Nellis, 1981; Fortov, 1995). Metallization of molecular hydrogen, though elusive in the diamond-anvil-cell experiments (Ruoff & Vanderbough, 1990; Mao, Hemley, & Hanfland, 1991; Hemley et al., 1996), was successfully demonstrated in experiments using compression through shock wave reverberation between electrically insulating sapphire (Al_2O_3) anvils (Weir, Mitchell, & Nellis, 1996; Da Silva et al., 1997), as we shall recapitulate in Sec. 7.2.

6.1.4 METALLIC HYDROGEN IN ASTROPHYSICAL OBJECTS

The giant planets such as Jupiter are thought to consist mostly of metallic hydrogen (Stevenson, 1982; Van Horn, 1991). The first-order MI transitions predict a discontinuous distribution and resistivity near the surface of Jupiter, implying a large enough magnetic Reynolds number to sustain the prominent magnetic activities (e.g., Kennel & Coroniti, 1977; Stevenson, 1982).

The release of latent heat associated with metal-to-insulator transitions through cooling may possibly account for a considerable fraction of its excess infrared luminosity (Aumann, Gillespie, & Low, 1969; Hubbard, 1980), as we shall revisit in Chap. 7.

Ferromagnetic and freezing transitions in the liquid-metallic hydrogen (Ichimaru, 2001) are important issues, not only in condensed-matter physics but in conjunction with the conspicuous magnetic phenomena in astrophysics, such as those associated with the origin of intense magnetization found in the degenerate stars (e.g., Chanmugam, 1992). Liquid-metallic hydrogen relevant to the ferromagnetic transitions may, for example, be expected in an outer layer of a hydrogen-rich white dwarf; we shall explore these in Chap. 8.

6.1.5 NUCLEAR REACTIONS

The rates of a nuclear process such as thermonuclear and pycnonuclear reactions are influenced significantly by the state or the phase that a dense matter may assume (e.g., Ichimaru, 1993). A huge enhancement of the reaction rates arising from internuclear Coulomb correlation in dense matter, albeit ineffective for the solar nuclear reactions or for the ICF experiments, provides a physical mechanism vital to supernovae.

Experimental and theoretical progress in ultrahigh-pressure metal physics may make a "supernova on the Earth" scheme utilizing enhanced pycnonuclear reactions in ultradense metallic hydrogen an attractive and possibly even realizable prospect for fusion studies; we shall take up on this subject in Chap. 9.

6.2 EQUATIONS OF STATE FOR HYDROGEN

Metallic fluid hydrogen consists of itinerant electrons (fermions with spin-½) and itinerant protons (classical, or fermions with spin-½) with strong electron-ion (e-i) coupling. Metallic solid hydrogen consists of itinerant electrons (fermions with spin-½) and a bcc array of protons (classical) with harmonic and anharmonic lattice vibrations with strong e-i coupling (e.g., Kitamura & Ichimaru, 1998; Ichimaru, 2004b).

Equations of state for those metallic fluid and solid hydrogen may be constituted through combinations of those presented in the previous chapter.

6.2.1 MOLECULAR FLUIDS

Among the potential functions describing the intermolecular forces, the Lennard-Jones potentials have been thought to be most desirable in light of accuracy, generality, and analyticity (e.g., Ceperley & Kalos, 1979).

For the intermolecular potentials between hydrogen molecules, we may use a Lennard-Jones potential,

$$V(r) = 4\varepsilon_m \left[\left(\frac{\sigma_m}{r} \right)^{12} - \left(\frac{\sigma_m}{r} \right)^6 \right]. \tag{6.3}$$

A statistical system interacting with a Lennard-Jones potential is represented by the three dimensionless parameters:

$$n_m^* = n_m \sigma_m^3, \tag{6.4a}$$

$$T_m^* = k_B T / \varepsilon_m, \tag{6.4b}$$

$$\Lambda_m^* = \frac{2\pi\hbar}{\sqrt{2m_p \varepsilon_m \sigma_m^2}}, \tag{6.4c}$$

with m_p and n_m denoting the mass of a proton and the number density of the molecules. These parameters characterize the density, temperature, and quantum-mechanical effects, respectively.

The specific Helmholtz free energy (per molecule in units of k_BT) of a molecular H_2 fluid is expressed as

$$f_{\text{mol-fl}} = f_m^{\text{id}} + f_{\text{HS}} + f_{\text{attr}} + f_{\text{rot}} + f_{\text{vib}} + \frac{E_{\text{H}_2}}{k_B T}. \tag{6.5}$$

The terms on the right-hand side consist of an ideal Bose gas, short-range repulsive interaction between molecular cores, attractive (dipolar) van der Waals forces, molecular rotation (roton), intramolecular vibration (vibron), and the ground-state energy of a H_2 molecule (E_{H2}) (Hansen & Verlet, 1969; Hansen, 1970; Hansen & McDonald, 1986; Kitamura & Ichimaru, 1998).

The analytic formula for f_m^{id}, applicable at any temperature and density, is expressed as

$$f_m^{\text{id}} = \frac{0.0327(T/T_c)^{5.09} f_0^{\text{cl}} + f_0^q}{1 + 0.0327(T/T_c)^{5.09}}. \quad (6.6)$$

In this expression,

$$f_0^{\text{cl}} = \ln\left\{ n_m^* \left[\frac{(\Lambda_m^*)^2}{2\pi T_m^*} \right]^{3/2} \right\} - 1 \quad (6.7a)$$

is the free energy of a classical Boltzmann gas and

$$f_0^q = -\frac{2}{3} \frac{J(5/2)}{J(3/2)} \left(\frac{T}{T_c} \right)^{3/2} \quad (6.7b)$$

is the corresponding expression for a degenerate Bose gas. The transition temperature between these two evaluations is expressed as

$$T_c = \frac{\varepsilon_m}{k_B} \left(\frac{\sqrt{2}\pi^2}{J(3/2)} \right)^{2/3} \frac{(\Lambda_m^*)^2 (n_m^*)^{2/3}}{4\pi^2} \quad (6.8)$$

where $J(n) = \Gamma(n) \cdot \zeta(n)$, with $\Gamma(n)$ and $\zeta(n)$ representing the gamma function and the zeta function (cf. Appendix IV).

6.2.2 MOLECULAR SOLIDS

The Helmholtz free energy of H_2 solid with the Lennard-Jones intermolecular potential is expressed as

$$f_{\text{mol-sol}} = f_{\text{coh}} + f_{\text{ph}} + f_{\text{rot}} + f_{\text{vib}} + \frac{E_{H_2}}{k_B T}. \quad (6.9)$$

Equations of state in the molecular-solid insulator phase of hydrogen consist of cohesive energy with a hexagonal-close-packed (hcp) structure (e.g., Ashcroft & Mermin, 1976),

$$f_{\text{coh}} = \frac{6.07(n_m^*)^4 - 14.45(n_m^*)^2}{T_m^*}, \quad (6.10)$$

lattice vibration (phonon), roton, vibron, and the ground-state energy of a H_2 molecule (E_{H_2}) (Hirschfelder, Curtis, & Bird, 1954; Kitamura & Ichimaru, 1998).

In addition, we take into account contributions of atomic hydrogen to the equations of state, which include the repulsive hard sphere, the attractive van der Waals, and the ground-state energy (E_H) of H atoms (Dargarno, 1967; Victor & Dargarno, 1970; Kitamura & Ichimaru, 1998).

6.3 PHASES OF HYDROGEN MATTER

The phase diagrams for the MI transitions in hydrogen matter may be determined through the explicit formulation of the equations of state in the metallic (solid, paramagnetic fluid, ferromagnetic fluid) phases as well as in the insulator (molecular solid, molecular fluid, atomic and molecular fluid) phases, as listed in the previous section. Thus, we consider a matter consisting of atomic, molecular, and ionized hydrogen, which may be characterized by the temperature T, the total number density of protons n_p, the degree of ionization $\langle Z \rangle$, and the degree of dissociation α_d. The number densities of ions, plasma electrons, neutral atoms, and molecules are then given by

$$n = \langle Z \rangle n_p, \quad n_e = n,$$

$$n_a = \alpha_d \left(1 - \langle Z \rangle\right) n_p, \quad n_m = \left(1 - \alpha_d\right)\left(1 - \langle Z \rangle\right) n_p / 2.$$

6.3.1 EQUATIONS OF STATE FOR THE FLUID PHASE

For a fluid phase, the total Helmholtz free energy, F_{tot}, may thus be expressed as a sum of molecular (6.11a), atomic (6.11b), metallic (6.11c), and intermolecular (6.11d) contributions in the following (Kitamura & Ichimaru, 1998):

$$f_{\text{tot}}\left(n_p, T; \langle Z \rangle, \alpha_d\right) \equiv \frac{F_{\text{tot}}}{V n_p k_B T}$$

$$= \frac{\left(1-\alpha_d\right)\left(1-\langle Z \rangle\right)}{2} \left(f_m^{\text{id}} + f_{\text{rot}} + f_{\text{vib}} + \frac{E_{H_2}}{k_B T} \right) \tag{6.11a}$$

$$+ \alpha_d \left(1 - \langle Z \rangle\right) \left(f_a^{\text{id}} + \frac{E_H}{k_B T} \right) \tag{6.11b}$$

$$+ \langle Z \rangle f_{\text{met-fl}}\left(\bar{\rho}_m, T\right) \tag{6.11c}$$

$$+ \frac{\left(1-\alpha_d\right)\left(1-\langle Z \rangle\right)}{2} \left(f_{\text{HS}} + f_{\text{attr}} \right). \tag{6.11d}$$

In this formulation, interaction between plasmas and neutral particles is taken into account through excluded-volume effects and changes in the levels of bound electrons: In the former effects, specific volume for the plasma particles (i.e., ions and electrons) is effectively reduced by the presence of neutral atoms and molecules, so that the term (6.11c) contains a normalized density, $\bar{\rho}_m = \rho_m / (1-\eta)$, where η is the packing fraction:

$$\eta = \frac{\pi}{6}\left(n_m d_m^3 + n_a d_a^3\right) \tag{6.12}$$

with d_a and d_m denoting the effective hard-sphere diameters of a hydrogen atom and molecule (Lebowitz & Rowlinson, 1964). Thus, the presence of neutral atoms has been effectively taken into account in (6.11d).

6.3.2 SHORT-RANGE SCREENING BY ELECTRONS

When metal and insulator phases coexist, the energy level of an electron bound in a molecule or in an atom in a dense plasma may be lowered, or may even disappear, owing to the screening action of plasma electrons (e.g., Ichimaru, 2004b). The extent to which such a modification may take place depends on the ratio between the Bohr radius and the short-range screening distance D_s of the plasma defined in terms of the dielectric response function $\varepsilon_e(k, 0)$ of the electrons as

$$\frac{1}{D_s} = \frac{2}{\pi}\int_0^\infty dk\left[1 - \frac{1}{\varepsilon_e(k,0)}\right]. \tag{6.13}$$

Thus, the ground-state energy of a hydrogen atom in a plasma may be expressed as

$$E_{\rm H}\,(\text{eV}) = -13.6 f_s(a_B / D_s),$$

where

$$f_s(x) = 1 - 1.9585x + 1.2172x^2 - 0.24900x^3 + 0.012973x^4. \tag{6.14}$$

This screening function has been obtained through the numerical solution to a Schrödinger equation for an electron in a Yukawa potential, $-(e^2/r)\exp(-r/D_s)$. As x increases from zero, the value of $f_s(x)$ decreases from $f_s(0) = 1$, meaning that an atomic or a molecular level is lowered; $f_s(x)$ vanishes at $x = 1.17$, where a bound state disappears.

Of significance in these connections is the essential difference between the two screening lengths defined by (6.13) and by the thermodynamics,

$$D_L^2 = \frac{1}{4\pi e^2}\left(\frac{\partial \mu}{\partial n}\right)_{T,V}, \tag{6.15}$$

with μ denoting the chemical potential of the screening electrons; the usual Debye–Hückel screening distance, λ_D in (1.14), is a version of this D_L.

Since D_L^2 has been defined in terms of the isothermal compressibility of the electrons, the latter quantity may take on a negative value at low densities (i.e., in the strong Coulomb coupling), when D_L would become an ill-defined quantity.

On the other hand, D_s in (6.13) remains a well-defined quantity, since one generally proves

$$1/\varepsilon_e(k,0) < 1 \tag{6.16}$$

from the causality requirement with a density–density response function (e.g., Ichimaru, 1982). Since D_s characterizes the short-range behavior of the screened Coulomb forces, it plays an essential part in calculating the rate of nuclear reactions in dense plasmas, as we shall recapitulate in Chap. 9.

A ground-state energy, E_{H2}, of a hydrogen molecule in plasma may likewise be calculated as

$$E_{H_2}\,(\mathrm{eV}) = -2\times 13.6 f_s\,(a_B\,/\,D_s) - 4.747.$$

The last numeral, 4.747, represents the dissociation energy in units of eV. We remark that this number does not contain the contribution of zero-point energies of the vibrons; the latter has been taken into account already in the term (6.11a).

6.4 COEXISTENCE CURVES AND THERMODYNAMICS

When the values of the mass density ρ_m and the temperature T are given, the chemical equilibrium of the system may be determined through the condition that the total free energy, (6.11a–d), be minimized with respect to $\langle Z \rangle$ and α_d. With the state of the matter so determined, we may calculate the values of the thermodynamic quantities in a standard way:

$$\text{pressure:}\quad P = -\left(\frac{\partial F}{\partial V}\right)_{T,\langle Z\rangle,\alpha_d}, \tag{6.17a}$$

$$\text{entropy:}\quad S = -\left(\frac{\partial F}{\partial T}\right)_{V,\langle Z\rangle,\alpha_d}, \tag{6.17b}$$

and so on (e.g., Landau & Lifshitz, 1969).

6.4.1 PHASE DIAGRAM AND COEXISTENCE CURVES

The phase diagram of hydrogen obtained through considerations of the equations of state for those various states is shown in Figure 6.1. The dotted curves are isobars at the designated pressure values. $\langle Z \rangle$ denotes the degree of ionization; α_d, the degree of atomic dissociation. C_{MI} and C_{GL} are the critical points associated with metal–insulator and gas–liquid transitions; T_{GLS}, the gas–liquid–solid triple point; T_{spf}, the triple point for the solid–paramagnetic–ferromagnetic phases; C_{mag}, the critical point for the ferromagnetic transitions. (Kitamura & Ichimaru, 1998)

Coexistence curves between the MI transitions are derived from the general conditions for the phase equilibrium (e.g., Landau & Lifshitz, 1969), that is,

$$P(\rho_M, T) = P(\rho_I, T), \tag{6.18a}$$

$$G(\rho_M, T) = G(\rho_I, T). \tag{6.18b}$$

The mass densities, ρ_M and ρ_I, are those of metallic and insulating hydrogen along the coexistence curves. The coexistence curves for the MI transitions in hydrogen

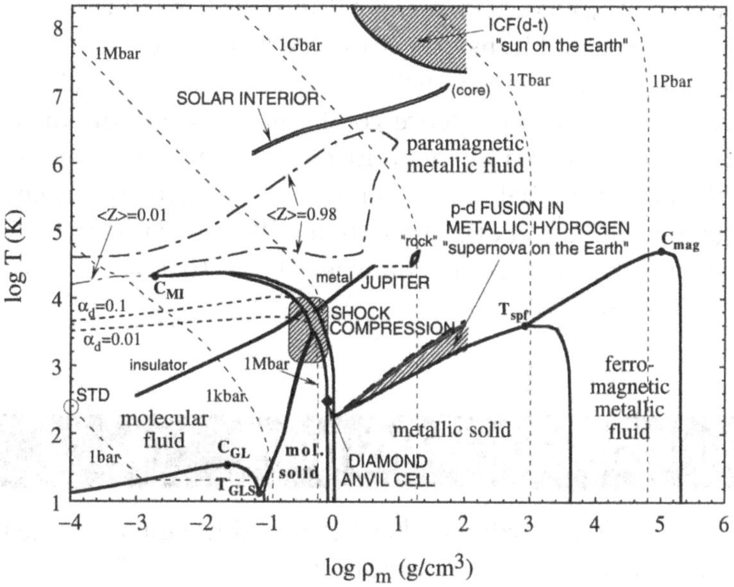

FIGURE 6.1 Phase diagram of hydrogen: The dotted curves are isobars at the designated pressure values. $\langle Z \rangle$ denotes the degree of ionization; α_d, the degree of atomic dissociation. C_{MI} and C_{GL} are the critical points associated with metal–insulator and gas–liquid transitions; T_{GLS}, the gas–liquid–solid triple point; T_{spf}, the triple point for the solid–paramagnetic–ferromagnetic phases; C_{msg}, the critical point for the ferromagnetic transitions. See the text for designations of examples cited for hydrogen matter in terrestrial laboratory and astrophysical settings (Kitamura & Ichimaru, 1998).

so obtained are also shown in Figure 6.1. Here, we clearly observe the first-order MI transitions exhibited in a high-density regime of hydrogen.

The diagrams have been constructed through explicit calculations of the equations of state for the metallic (solid, paramagnetic fluid, ferromagnetic fluid, partially ionized atomic and molecular fluid) phases as well as in the insulator (molecular solid, molecular fluid, atomic and molecular fluid) phases.

In Figure 6.1, we also exhibit approximate parameter domains for the hydrogen matter appropriate to the solar interior, the Jovian interior, inertial-confinement fusion (ICF) researches, ultrahigh-pressure metal experiments (shock or diamond-anvil-cell compression), proton-deuteron (p-d) fusion in ultradense metallic hydrogen, and hydrogen at standard (STD) conditions. (The "rock" shown here for the core of Jupiter is not of hydrogen, as explained in Figure 1.1.)

6.4.2 THERMODYNAMICS ACROSS THE MI TRANSITIONS

Table 6.1 lists the thermodynamic characteristics of the MI transitions as functions of the pressure, down to the critical point C_{MI}, where $\log P$ (bar) = 3.477; $\log T$ (K) = 4.320; $\log \rho_m \left(g/cm^3\right) = -2.721$ (Kitamura & Ichimaru, 1998). We find that the increment of the specific entropy, $s \equiv S/Vnk_B$, between metal and insulator phases, $\Delta s_{MI} (\equiv s_M - s_I) > 0$; hence, the insulator-to-metal transition is an *endothermic* process.

Summarizing the theoretical predictions in Figure 6.1 and Table 6.1 for the MI transitions, we note: In the coexistence conditions between the metal and insulator phases, all the thermodynamic quantities except for the pressure, the temperature, and the chemical potential are discontinuous. The density, the entropy, and the enthalpy are greater in the metallic phase than in the insulator phase. These discontinuities vanish at the critical point C_{MI}, where $\log P$ (bar) = 3.477; $\log T$ (K) = 4.320; $\log \rho_m \left(g/cm^3\right) = -2.721$.

6.5 METAL–INSULATOR TRANSITIONS

The MI transitions in dense hydrogen have attracted the interest of many investigators since the pioneering work of Wigner and Huntington (1935).

Friedel and Ashcroft (1977) carried out approximate electron band calculations that predicted a band crossing at $\rho_m = 0.82\ g/cm^3$. Analogous calculations were performed by Chacham and Louie (1991) for the band gap of solid molecular hydrogen in the *hcp* structure, which predicted that the orientationally ordered phase undergoes metallization due to an indirect band overlap at $\rho_m = 0.8\ g/cm^3$ and an orientationally disordered phase at $\rho_m = 1.06\ g/cm^3$.

In conjunction with shock-compressed states determined by the Hugoniot relations, Ross, Ree, & Young (1983) and Holmes, Ree, & Young. (1995) obtained so-called

TABLE 6.1 Thermodynamic Quantities for Dense Hydrogen along the Metal–Insulator Coexistence Curves as Functions of Pressure.

$\log P$ (bar)	$\log T$ (K)	$\log\rho_I$ (g / cm^3)	$\log\rho_M$ (g / cm^3)	s_I	s_M	Δs_{MI}
6.385	1.000	−0.0738	0.0284	1.48E−7	5.04E−3	5.04E−3
6.385	1.301	−0.0738	0.0284	1.20E−6	1.02E−2	1.02E−2
6.385	1.699	−0.0738	0.0284	1.90E−5	2.81E−2	2.81E−2
6.385	2.000	−0.0740	0.0283	2.49E−3	7.35E−2	7.10E−2
6.384	2.267	−0.0745	0.0281	0.130	0.200	7.00E−2
6.384	2.267	−0.0745	0.0264	0.130	3.510	3.380
6.368	2.477	−0.0857	0.0216	0.832	2.298	1.466
6.347	2.699	−0.0973	0.0136	0.938	2.769	1.831
6.286	3.000	−0.131	−0.0108	1.564	3.938	2.374
6.215	3.176	−0.169	−0.0394	2.271	4.823	2.552
6.140	3.301	−0.209	−0.0698	3.022	5.563	2.541
6.000	3.473	−0.288	−0.128	4.399	6.657	2.258
5.981	3.493	−0.299	−0.136	4.585	6.797	2.212
5.981	3.493	−0.327	−0.136	5.487	6.797	1.310
5.924	3.602	−0.356	−0.168	6.086	7.453	1.367
5.852	3.699	−0.401	−0.207	6.781	8.060	1.279
5.719	3.845	−0.491	−0.288	7.967	9.022	1.055
5.541	4.000	−0.624	−0.418	9.254	10.146	0.892
5.000	4.234	−1.087	−0.884	11.516	12.715	1.199
4.754	4.301	−1.357	−1.079	12.397	13.656	1.259
4.531	4.365	−1.648	−1.500	13.283	15.897	2.614
4.126	4.352	−2.061	−2.221	14.135	18.763	4.628
4.000	4.342	−2.186	−2.337	14.365	19.112	4.747
3.754	4.327	−2.433	−2.526	14.904	19.602	4.698
3.477	4.320	−2.721	−2.721	20.385	20.385	0.000

Note: "Ex" Means "10^X"

shock equations of state with the aid of adopted model potentials and predicted a transition to a *bcc* metal above 5 Mbar.

Equations of state in dense hydrogen were investigated with model intermolecular potentials to predict the conditions for pressure ionization (Brovman, Kagan, & Kholas, 1972; Ebeling et al., 1991; Saumon, Chabrier, & Van Horn, 1995).

Quantum Monte Carlo simulations were performed for MI transitions in the ground state (Ceperley & Alder, 1987) as well as at elevated temperatures (Magro et al., 1996).

Recent progress in the computer simulation studies on the properties of hydrogen and helium under extreme conditions has been extensively reviewed (McMahon et al., 2012).

An insulator-to-metal transition (i.e., metallization) proceeds in the direction of decreasing the chemical potential, accompanied by an increase of the temperature and/or decrease of the pressure. It is thus an endothermic process, analogous to our familiar vaporization and melting transitions, as remarked earlier.

When a change of states takes place, the enthalpy W of the system varies from the initial to the final by an amount,

$$\Delta W = \int_{\text{initial}}^{\text{final}} VdP + \int_{\text{initial}}^{\text{final}} TdS. \tag{6.19}$$

The first term on the right-hand side describes a hydro-mechanical compression; the second term, a thermal process that involves the release of a *latent heat.*

Looking from somewhat different directions, we note: When hydrogen in an insulator phase is compressed to a state of high density such that average inter-particle spacing between protons becomes comparable to or less than the orbital radii of the bound electrons, which are on the order of the Bohr radius, electrons begin to assume *itinerant* states due to overlapping of wave functions between adjacent electrons. It is a pressure ionization, mentioned earlier, which takes place instantaneously as in an electric breakdown.

Another class of metallization may take place when the temperature is raised above the atomic or molecular binding energies of the electron. It is thermal ionization, whose degree of ionization changes continuously as the temperature varies.

7

TRANSPORT PROCESSES

A dense plasma material may be modeled as a two-component plasma (TCP) constituted by electrons and ions. In the TCP, attractive interaction between electrons and ions, an essential ingredient for the formation of atoms, brings about novel features in that the strong correlations between electrons and ions are taken into account on an equal footing with the atomic and molecular processes in such a condensed environment. These dense-plasma effects may also influence the atomic levels themselves. The strong electron-ion (e-i) coupling thus opens up new dimensions in condensed plasma physics, whereby interplay with the atomic and molecular physics plays a central part. We may recall these features observed already in the metal–insulator (MI) transitions.

In this chapter, we first consider the electric and thermal resistivity in such dense plasmas. The results are then applied to the explanation of an ultrahigh-pressure metal physics experiment as well as to the elucidation of Jovian excess infrared luminosity.

7.1 ELECTRIC AND THERMAL RESISTIVITY

Electric and thermal resistivity arises as a consequence of scattering between electrons and ions in plasmas. A proper treatment of such e-i interactions is quite essential, as the resistivity would diverge in a classical treatment of scattering at short distances.

Hubbard and Lampe (1969) investigated thermal conduction by electrons in dense stellar matter through a Chapman–Enskog solution to the quantum-mechanical transport equation in weak Coulomb coupling. On the basis of a quantum-mechanical theory for the current–current correlation functions, Boercker, Rogers, and DeWitt (1982) obtained an expression for electric resistivity, where electrons were treated semi-classically in their numerical calculations.

In an earlier investigation, Tanaka, Yan, and Ichimaru (1990) calculated the resistivity of dense hydrogen plasmas through solutions to quantum-statistical transport equations for the electrons (e.g., Ichimaru, 2004a,b). It was shown in particular that quantum diffraction of electrons in the vicinity of ions, called *incipient Rydberg state* (IRS) effects, plays a major part in the Coulomb resistivity for dense plasmas close to the MI transitions treated in the preceding chapter.

In the treatment of the MI transitions, Kitamura and Ichimaru (1995) extended the results to the cases including high-Z TCP. They then performed additionally the partial-wave analyses in the ion-sphere model on the Z-dependent effects for the scattering cross-sections.

We may also note a related effect in the short-range screening by electrons, which lowers the energy level of an electron bound in a molecule or in an atom in dense plasmas. Such an effect may even act to eliminate the level, owing to the strong action of screening electrons, as we have seen in Sec. 6.3.2.

7.1.1 PARAMETERIZED FORMULAE

We begin with the parameterized formulae of the resistivity in fully ionized plasmas with the ionic charge number Z, the number densities n_i, and n_e of the ions and the electrons. Electric and thermal resistivity, ρ_E and ρ_T, arising from e-i scattering are expressed as (Kitamura & Ichimaru, 1995)

$$\rho_E = \frac{8}{3}\sqrt{\frac{\pi}{2}}\frac{Z^2 e^2 \sqrt{m n_i}}{n_e (k_B T)^{3/2}} L_E, \tag{7.1}$$

$$\rho_T = \frac{C_P^{(0)}}{C_P}\frac{52\sqrt{2\pi}}{75}\frac{Z^2 e^4 \sqrt{m n_i}}{n_e k_B (k_B T)^{5/2}} L_T, \tag{7.2}$$

where C_P and $C_P^{(0)}$ are the specific heat (per electron) at constant pressure for the plasma and for the ideal-gas electrons, respectively. In dense plasmas, due to the presence of strongly coupled ions, the values of $C_P^{(0)} / C_P$ can be significantly smaller than unity; this effect may thus act to enhance the thermal transport.

7.1.2 GENERALIZED COULOMB LOGARITHMS

Formulae (7.1) and (7.2) define the *generalized Coulomb logarithms*, L_E and L_T. In the classical ($\theta \gg 1$) and weak-coupling ($\Gamma \ll 1$) regime, both L_E and L_T approach the Debye–Hückel limiting values,

$$L_0 = -\frac{1}{2}\ln\varsigma - \frac{1}{2}\left[\gamma + \frac{1}{Z}\ln(Z+1)\right] + O(\varsigma), \tag{7.3}$$

where

$$\varsigma = \frac{\hbar^2 (Z+1) k_e^2}{8 m k_B T},$$

and $\gamma = 0.57721\ldots$ is Euler's constant. This formula has been derived by Kivelson and DuBois (1964) with the aid of the quantum-mechanical Lenard–Balescu–Guernsey equation (e.g., Ichimaru, 2004a).

7.1.3 SCREENED POTENTIALS

In the degenerate ($\theta \ll 1$) and strong-coupling ($\Gamma \gg 1$) regime, the inter-particle correlations are described by the ion-sphere model of Sec. 1.3.3, in which one considers an ion surrounded by a uniform electronic charge sphere of the ion-sphere radius a, as depicted in Figure 1.5. In this model, the potential of scattering around an ion is expressed as

$$U(r) = \begin{cases} Ze^2\left(-\dfrac{1}{r} + \dfrac{3}{2a} - \dfrac{r^2}{2a^3}\right) & \text{for } r \le a, \\ 0 & \text{for } r > a, \end{cases} \tag{7.4}$$

where the position of the ion is set at the origin, $\mathbf{r} = 0$.

The resistivity is proportional to the transport cross-section, $Q_{\mathrm{m}}(k_{\mathrm{F}})$, for the electrons with wave number $k_{\mathrm{F}} = (3\pi^2 n_{\mathrm{e}})^{1/3}$ (e.g., Landau & Lifshitz, 1965). This quantity may be evaluated from the phase shifts, obtained through numerical solutions to the Schrödinger equation with the potential $U(r)$.

The Born approximation is applicable for $E_{\mathrm{F}} > Z^2$ Ry. In this regime, the values of $Q_{\mathrm{m}}(k_{\mathrm{F}})$ obtained through the phase shift analyses, in fact, show good agreement with the results in the Born approximation (Kitamura & Ichimaru, 1995), which can be expressed in a fitting formula,

$$Q_m^{\mathrm{Born}}(k_F) = 1.14 a^2 r_s^2 Z^{8/3} \exp(-1.47 Z^{1/3}). \tag{7.5}$$

This formula is applicable for $Z \le 26$.

On the basis of the results, (7.3) and (7.5), applicable to both ends of the limit, we may express the Coulomb logarithms as

$$L_{E(T)} = \frac{1}{2}\ln\left[1 + \alpha_{E(T)}\left(\frac{1}{\varsigma_{\mathrm{DH}}} + \tanh\frac{1}{\varsigma_{\mathrm{Born}}}\right)\right] \times \left\{1 + A_{E(T)} x_b^2 \exp\left(-C r_s^D\right) + B_{E(T)} x_b^{10} \exp\left(-5 C r_s^D\right)\right\} \tag{7.6}$$

where $\alpha_{\mathrm{E}} = 1$, $\alpha_{\mathrm{T}} = 75/13\pi^2 = 0.5845\ldots$, and

$$\varsigma_{\mathrm{DH}} = \frac{(2/3)^{2/3} \exp\gamma}{\pi}(Z+1)^{1+1/Z}\frac{r_s}{\theta^2}, \tag{7.7a}$$

$$\varsigma_{\mathrm{Born}} = \frac{\exp(1.47 Z^{1/3})}{K\theta^{3/2} Z^{4/3}}. \tag{7.7b}$$

In these formulae, $A_E = 0.42$, $B_E = 0.063$, $A_T = 0.38$, $B_T = 0.049$, $C = 6\times10^{-4}$, $D = 2$, and $K = 2.5$, which have been determined through fit to computed results for hydrogen plasmas at $0.01 \le \theta \le 10$, $0.05 \le \Gamma \le 43.441$, and $x_b \le 1.5$ (Tanaka, Yan, & Ichimaru, 1990).

7.1.4 THE IRS PARAMETER

Here and in (7.6),

$$x_b = \left\{ r_s \tanh\left[\hbar \left(\frac{2\pi}{mk_BT} \right)^{1/2} n^{1/3} \right] \right\}^{1/2} \tag{7.8}$$

is a dimensionless IRS parameter, proportional linearly to e, so that we observe

$$x_b^4 = \begin{cases} \left(\frac{9\pi}{4} \right)^{2/3} \frac{\text{Ry}}{E_F} & (\theta \ll 1) \\ (36\pi)^{1/3} \frac{\text{Ry}}{k_BT} & (\theta \gg 1) \end{cases}.$$

Thus, the quantity x_b^4 measures the strength of Coulomb coupling between electrons and ions relevant to the MI transitions, as it represents a ratio between a binding energy of a hydrogen atom and a kinetic energy of a free electron,.

The term inside the braces in the formula (7.6), a steeply increasing function of x_b, describes the enhancement of scattering due to the strong e-i Coulomb coupling. In the low-density ($r_s \gg 1$) limit, however, the IRS effects should vanish since the probability of finding electrons within a Bohr radius of an atom is small; the factor $\exp(-Cr_s^D)$ accounts for such an effect.

Those analytic formulae retain the following features: In the classical ($\theta \gg 1$) and weak-coupling ($\Gamma \ll 1$) case, (7.6) reproduces the Debye–Hückel result (7.3) since $\zeta_{DH} \ll 1$ and $\zeta_{Born} \ll 1$. In the degenerate ($\theta \ll 1$) and strong-coupling ($\Gamma \gg 1$) case, $L_E \approx \alpha_E/2\zeta_{Born}$ and $L_T \approx \alpha_T/2\zeta_{Born}$, since $1 \ll \zeta_{Born} < \zeta_{DH}$. The transport cross-section $Q_m(k_F)$ obtained from ρ_E via a Drude formula,

$$Q_m(k_F) = \frac{n_e e^2 \rho_E}{\hbar n_i k_F}, \tag{7.9}$$

is thus proportional to $Q_m^{\text{Born}}(k_F)$ of (7.5); the Wiedemann–Franz relation (e.g., Ichimaru, 2004a),

$$\frac{\rho_E}{\rho_T} = \frac{\pi^2 k_B^2 T}{3e^2} \tag{7.10}$$

is satisfied.

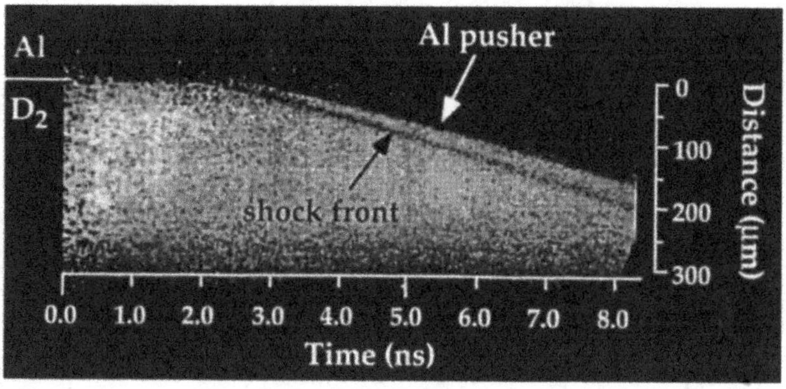

L.B. Da Silva (Lawrence Livermore Lab.)

FIGURE 7.1 Shock-compression experiment: A time-resolved side-on radiograph of a laser-shocked D_2 cell. Position of the pusher (Al) and the evolving shock front are measured as functions of time. After Da Silva et al. (1997).

7.2 ULTRAHIGH-PRESSURE METAL PHYSICS EXPERIMENTS

Weir, Mitchell, and Nellis (1996) compressed molecular fluid hydrogen to pressures ranging 0.93–1.80 Mbar by shock wave reverberation between insulating Al_2O_s anvils (Figure 7.1) and thereby measured the pressure and the electric resistivity attained in seven runs of hydrogen compression/metallization experiments with the data, as exhibited in Figure 7.2. These authors then interpreted the experimental data in terms of a *continuous* transition from semiconducting to metallic *diatomic* fluids associated with a closure of a semiconductor band gap E_g, near 1.4 Mbar.

Such an interpretation, however, contradicts against any of the theoretical predictions (Wigner & Huntington, 1935; Ceperley & Alder, 1987; Saumon, Chabrier, & Van Horn, 1995; Magro et al., 1996), which would foresee *first-order* insulator-to-metal transitions from molecular to *monatomic* hydrogen. On the basis of the equations of state described in Chap. 6 and the electric resistivity of dense hydrogen near the MI transitions in Sec. 7.1, it has been shown that those experimental results can be interpreted consistently with the phase diagram of hydrogen exhibiting the first-order MI transitions in Figure 6.1 as well as in Table 6.1 (Kitamura & Ichimaru, 1998).

7.2.1 INTERPRETING THE EXPERIMENTS

In interpreting the experiments, it is useful first of all to examine the relevant time scales in the compression and metallization processes involved. Let the thickness of the compressed hydrogen be ξ. Typical values of ξ in the experiments (see Figure 7.1) are on the order of 100 μm. A hydrodynamic or compression time, τ_H, estimated as ξ divided by a sound velocity may take on a value on the order of 10 ns.

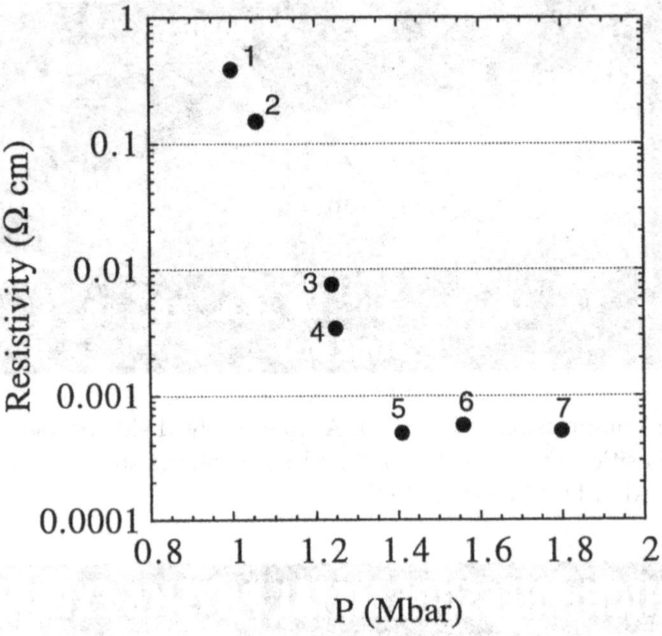

FIGURE 7.2 Values of the pressure and the electric resistivity measured in seven runs of hydrogen compression/metallization experiments. After Weir, Mitchell, and Nellis (1996).

An electronic or ionization time, τ_E, estimated as ξ divided by the Fermi velocity, that is, the Planck constant times the Fermi wave number k_F, may take on a value of about 60 ps. Since $\tau_E \ll \tau_H$, we remark that metallization develops "instantaneously" in a manner analogous to electric breakdown.

In a case where the metallization is partial, the heavy (metal) component ρ_{pl}, deposited uniformly in the volume, should in principle phase separate from the light (insulator) component ρ_{ins} by gravity. Since a maximally possible gravitational displacement may be estimated at its free-fall value, $g\tau_H^2 \approx 1$ fm, where $g \approx 980$ cm/s^2, we find the phase separation predicted in Figure 6.1 or in Table 6.1 cannot actually materialize in the experiments; the hydrogen matter may remain in a uniform (transient) state with the mass density, $\rho_m = \rho_{pl} + \rho_{ins}$.

7.2.2 COMPRESSION AND METALLIZATION

In light of these quantitative examinations, we may portray the shock-metallization experiments in two sequential stages:

Compression: In a typical experiment, hydrogen (with a total mass of about 20 mg) is compressed from the state ($P \approx 1$ bar, $T \approx 20$ K) through reverberating shock imparted by an Al-Al_2O_3 impactor. An impactor with a mass of 2–3 g and a velocity of 5–6 μm/ns, which are typical experimental parameters, carries a kinetic energy on the order of

50 kJ. The hydrogen pressure takes on its maximal value when the state of hydrogen reaches the insulator side of the MI transitions; the pressures measured in the experiments ($P = 1.0$–1.5 Mbar) correspond to these maximal values.

The time for such a compression is several times τ_H. Here, change in the enthalpy stems mostly from the hydro-mechanical contribution in (6.19).

Metallization: As metallization progresses in a time τ_E subsequent to that maximal-pressure state, the entropy, the temperature, and the enthalpy of the hydrogen increase. The efficiency of the metallization involves delicate matching between these changes and the dynamic conditions of the compression cell.

We observe in particular that the entropy increment through metallization is considerably large in Table 6.1 and that the pressure tends to decrease due to a temperature increase by metallization in the experiments. Hence, the endothermic increment ΔW (>0) must have come from "thermal" contribution (i.e., the integral of TdS, with the change in the entropy dS stemming from redistribution in the microscopic electronic states) rather than from a "hydrodynamic" contribution (i.e., the integral of VdP, which is negative).

A shock (and a diamond-anvil) compression, being hydrodynamic in nature, does not provide optimum conditions for such metallization; the enthalpy has increased only by ~0.3 kJ through the metallization experiments. We remark that a shock analysis of the Rankine–Hugoniot type is not applicable to such a non-hydrodynamic process in a short time-scale of τ_E.

Thus, to achieve efficient metallization, an additional physical mechanism such as injection of intense, ultra-short laser pulses into compressed hydrogen would have to be considered. We shall come back to this subject later, in Sec. 9.5.

7.2.3 EXAMINING THE DATA

The resistivity measured in a given experiment provides a measure of metallization attained. In all the experimental cases analyzed here, the metalized hydrogen is in a state of uniform, partially ionized matter consisting of electrons, protons, and molecules.

We estimate that the contributions of molecular scattering fall within the uncertainties (25%–50%) of the measured resistivity; the dominant cause of the resistivity is the Coulomb scattering between electrons and ions. The partial mass density ρ_{pl} of the metalized hydrogen may thus be assessed from the measured resistivity by the formula (7.1) of the Coulomb resistivity for a fully ionized hydrogen plasma; the results are shown in Figure 7.3.

We find that ρ_{pl} can be determined almost independent of the temperature since the carrier electrons are degenerate. The strong density dependence, proportional to $\rho_{pl}^{-2.5}$, of the theoretical resistivity reveals the significance of accounting for the strong e-i coupling beyond the Born approximation, a feature missing in Weir, Mitchell, and Nellis (1996).

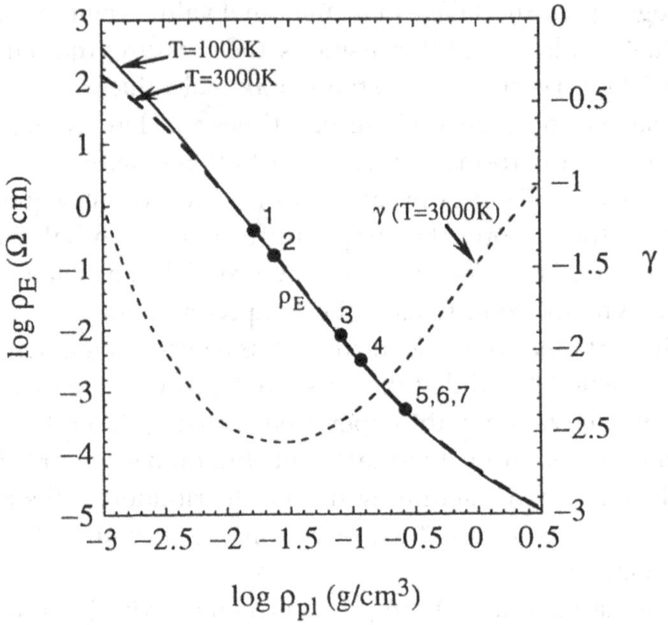

FIGURE 7.3 Electric resistivity vs. carrier density in hydrogen plasmas, calculated at $T = 1000$ K and 3000 K, by the theoretical formula (7.1). The dotted curve represents $\gamma \equiv d\ln\rho_E / d\ln\rho_{pl}$ computed at $T = 3000$ K. The solid circles refer to the experimentally determined resistivity (cf. Figure 7.2) for runs 1–7 (Weir, Mitchell, & Nellis, 1996) plotted on the theoretical curves at $T = 3000$ K.

7.2.4 THE FIRST-ORDER MI TRANSITIONS JUSTIFIED

The density ρ_{pl} is then connected to the pressure in the final state by recognizing the (negative) Coulomb pressure—Eq. (5.19c)—in dense metalized hydrogen is the major source of the decrease in the pressure. This decrease may thus be set equal to

$$P_{\text{met}} - P_{\text{ins}} = -\alpha e^2 \left(\frac{4\pi}{3}\right)^{1/3} \left(\frac{\rho_{\text{pl}}}{m_{\text{p}}}\right)^{4/3}, \tag{7.11}$$

which then determines the pressure P_{met} in the metalized state.

The parameter α would depend on an efficiency of energy transfer from an Al-Al_2O_3 impactor to the H_2 system through metallization. Since information on α is not available, we have chosen $\alpha = 0.02$ as its magnitude, which ensures the endothermic nature of metallization (i.e., $\Delta W > 0$) for each of the seven runs of the experiments.

Finally, the mass density can be obtained through the relation,

$$\rho_{\text{ins}} = \begin{cases} \rho_I(P_{\text{met}})\left[1 - \dfrac{\rho_{\text{pl}}}{\rho_M(P_{\text{met}})}\right], & \text{for } \rho_{\text{pl}} \le \rho_M(P_{\text{met}}), \\ 0, & \text{for } \rho_{\text{pl}} > \rho_M(P_{\text{met}}). \end{cases} \tag{7.12}$$

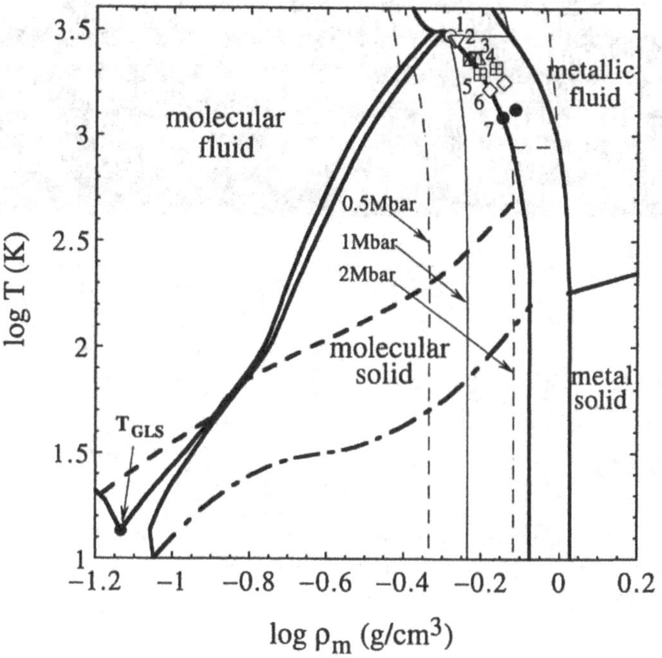

FIGURE 7.4 Plots of the initial (insulator) and final (metalized) states assessed for the experiments (Weir, Mitchell, & Nellis, 1996). The dashed and chain curves represent the theoretical adiabats starting from the initial conditions: $P = 1$ bar, $T = 20$ K (molecular fluid); and $P = 1$ bar, $T = 10$ K (molecular solid), respectively (Kitamura & Ichimaru, 1998).

Figure 7.4 depicts the maximum-pressure states and the final metalized states so determined for the runs 1–7 of the experiments (Weir, Mitchell, & Nellis, 1996); the corresponding thermodynamic quantities are listed in Table 7.1 for the runs 1, 3, 5, 7.

For comparison, also plotted in Figure 7.4 are the theoretical adiabats starting from the initial conditions: $P = 1$ bar, $T = 20$ K (molecular fluid); $P = 1$ bar, $T = 10$ K (molecular solid). The experiments have been conducted under the molecular-fluid initial conditions and we observe the adiabat connects them to a state with $P = 2.2$ Mbar, $T = 530$ K, $\rho_m = 0.79$ g/cm^3 on the insulator side of the MI coexistence curves. We observe that each of the temperatures at the experimental points consistently stay above 530 K at the end point of the adiabat; the excess then measures the extent to which extra heating is incurred by non-adiabaticity during the shock compression.

It has thus been shown that all the experimental data displayed in Figure 7.2 can be placed consistently with the phase diagram (Figure 6.1) of hydrogen exhibiting the first-order MI transitions.

7.3 JOVIAN INTERIORS AND EXCESS INFRARED LUMINOSITY

The interiors of giant planets (Jupiter, Saturn, Uranus, Neptune) offer important objects of study in the condensed-matter physics of hydrogen. Models for the internal

TABLE 7.1 Thermodynamic Quantities Estimated at the Initial (Insulator) States and Their Increments in the Final (Metalized) States, in the Runs, 1, 3, 5, 7, for Hydrogen of the LLNL Experiments (Weir, Mitchell, & Nellis, 1996). Chemical Potential (μ) and Specific Enthalpy (*w*) Are in Temperature Units per Mass of a Hydrogen Atom; Specific Entropy (*s*) in Units of k_B Per Mass of a Hydrogen Atom

Run	1		3		5		7	
P (Mbar)	1.0	−0.002	1.24	−0.013	1.41	−0.067	1.80	−0.067
T (10^3 K)	2.97	0.01	2.31	0.03	1.96	0.12	1.22	0.11
μ (10^3 K)	−162.6	−0.11	−154.7	−0.38	−150.4	−1.48	−141.3	−1.34
s	4.40	0.08	3.48	0.37	2.96	1.19	1.87	1.05
w (10^3 K)	−149.5	0.18	−146.6	0.59	−144.6	1.34	−139.0	0.27
ρ (g/cm^3)	0.516	0.004	0.583	0.020	0.623	0.062	0.712	0.056
ρ_{pl}/ρ_m	0	0.032	0	0.135	0	0.402	0	0.359

Note: Chemical potential (μ) and specific enthalpy (w) are in temperature units per mass of a hydrogen atom; specific entropy (s) in units of kB per mass of a hydrogen atom.

Source: Weir, Mitchell, & Nellis, 1996

structures of these planets were proposed on the bases of the thermodynamic and transport properties of the interiors, the surfaces, and the atmosphere coupled with observational data such as gravitational harmonics (Stevenson, 1982; Hubbard & Marley, 1989).

7.3.1 STRUCTURE OF JUPITER

Typically, Jupiter has the radius $R_J \approx 7\times10^4$ km, some 11 times that of the Earth and approximately 1/10 of the solar radius, and the mass $M_J \approx 1.90\times10^{30}$ g, some 300 times that of the Earth and approximately 1/1000 of the solar mass. Model ranges of the mass density, the temperature, and the pressure of its interiors, consisting of the central "rock," the "metal" hydrogen with a few percent (in molar fraction) admixture of helium, are displayed in Figure 1.1.

As we noted in Sec. 1.1.1, the visible luminosity of the bright planet Jupiter, in fact, originates from solar radiation reflected from its surface, with albedo at 0.35. Jupiter has been known to emit radiation energy in the infrared range, approximately 2.7 times as intense as the total amount of radiation that it receives from the Sun. By observation through terrestrial atmospheric transmission windows at 8–14 μm (Menzel, Coblentz, & Lampland, 1926) and 17.5–25 μm (Low, 1966), Jupiter has been known to be an unexpectedly bright infrared radiator. This feature has been reconfirmed quantitatively by a telescope airborne at an altitude of 15 km and through flyby measurements with *Pioneer 10* and *Pioneer 11* spacecraft. For Jupiter, the effective surface temperature determined from integrated infrared power over 8 to 300 μm was 129 ± 4 K, while the surface temperature calculated from equilibration with the absorbed solar radiation was 109.4 K (Hubbard, 1980); the balance needs to be accounted for by internal power generation; hence, the issue of *excess infrared luminosity* of Jupiter.

In these connections, we particularly note the precise measurements of the Jovian gravitational field made recently by NASA's *Juno* spacecraft (Fortney, 2018). Adriani et al. (2018) reported observation of clusters of cyclones encircling Jupiter's poles. Iess et al. (2018) reported measurement of Jupiter's asymmetric gravity field. Kaspi et al. (2018) reported observation of Jupiter's atmospheric jet streams extending thousands of kilometers deep. Guillot et al. (2018) reported on a suppression of differential rotation in Jupiter's deep interior.

7.3.2 ORIGINS OF THE EXCESS LUMINOSITY

Over the evolution period ($\sim 4.6 \times 10^9$ yr) of Jupiter, the estimated excess luminosity, $L_{ex} \approx 4.6 \times 10^{17}$ W, would amount to the total released energy of approximately 6.5×10^{34} J. To account for the source of such an excess infrared luminosity, theoretical models such as "adiabatic cooling" (Hubbard, 1968; Graboske et al., 1975; Stevenson & Salpeter, 1976) and "gravitational unmixing" (Stevenson & Salpeter, 1976; Smoluchowski, 1967) have been considered. Here, we apply the phase diagram (Figure 6.1) and the electric-resistivity calculations of Sec. 7.1 to the issues of Jovian internal structure and luminosity.

7.3.3 THE MI TRANSITIONS AND LUMINOSITY

The first-order MI transitions in hydrogen predict the existence of a boundary layer near the surface of Jupiter across which the mass density and resistivity change discontinuously (Figure 1.1). Assuming the temperature of the boundary layer at 6.5×10^3 K (Van Horn, 1991; Kitamura & Ichimaru, 1998), we calculate the density in the outer insulator side to be 0.34 g/cm^3 and that of the inner metal side to be 0.54 g/cm^3.

Some 4.6 billion years ago, when our solar system was formed, the temperature was so high that hydrogen in Jupiter was in the ionized metallic state. The outer insulator side has then been formed over the evolution period through metal-to-insulator transitions.

The mass of the outer (molecular hydrogen) layer so formed was estimated to be $M_{ins} \approx 0.1 \text{x} M_J$ (Hubbard & Marley, 1989). Hence, the total amount of latent heat released through the MI transitions, that is, the thermal contributions in (6.19), is $(M_{ins} / m_p) \Delta s_{MI} k_B T_{MI}$, where the entropy increment $\Delta s_{MI} = 1.1$ and $T_{MI} = 6.5 \times 10^3$ K (Table 6.1). These approximate calculations may thus indicate that the total latent heat would amount to 1.1×10^{34} J, possibly accounting for about 1/6 of the energy in the excess infrared luminosity.

A final solution to these issues of transformation and transfer of energy, however, should await further investigations into the internal structures and evolution of Jupiter.

8

STELLAR AND PLANETARY MAGNETISM

Astrophysical magnetic phenomena include those related to degenerate stars (e.g., Chanmugam, 1992), solar flares (e.g., Parker, 1979; Tsuneta, 1995), and giant planets (e.g., Stevenson, 1982). Surface magnetic fields of magnetic white dwarfs range 10^6–10^9 gauss. Average strengths of magnetic activities in the solar chromosphere are on the order of 50 gauss.

Hydrogen is the major constituent in astronomical objects such as stars and giant planets. Stellar and planetary magnetism may thus be strongly influenced by the states and phase transformations in hydrogen matter such as metallization and magnetization.

8.1 JOVIAN MAGNETIC ACTIVITIES

The structure of Jupiter was illustrated in Figure 1.1 along with the phase diagrams of hydrogen exhibited in Figure 6.1. It is an astronomical object with a radius of $R_J \approx 7.14 \times 10^4$ km and a total mass of $M_J \approx 1.99 \times 10^{33}$ g.

The dominant field contribution of the planet Jupiter for the external observer is the dipole of magnitude 4.2 gauss$\cdot R_J^3$ and a tilt of ~10° to the rotation axis (Smith, Davis Jr., & Jones 1976).

Closer to the planet, however, the multipole contributions are so large that an additional dipole term at a depth of $\sim 2 \times 10^4$ km appears to be implied (Elphic & Russel, 1978).

8.1.1 METALLIC HYDROGEN IN JUPITER

The first-order metal–insulator (MI) transitions in hydrogen may predict the existence of a boundary layer inside Jupiter across which the mass density and the resistivity change discontinuously, as Fig. 1.1 implies.

With the estimates of Jovian parameters across the MI boundary in Figure 6.1 and Table 6.1, we calculate the electric resistivity ρ_E of the metallic hydrogen inside the MI discontinuity to be $1.37 \times 10^{-4}\ \Omega$ cm (Kitamura & Ichimaru, 1995), only about 140 times greater than that of copper.

8.1.2 MAGNETIC REYNOLDS NUMBER

Highly conductive liquid-metallic hydrogen in motion is capable of distorting and amplifying the magnetic field configurations of stars and planets. Hydromagnetic motion of the magnetic fields in a plasma may be described by the induction equation,

$$\frac{\partial \mathbf{B}}{\partial t} = \nabla \times (\mathbf{v} \times \mathbf{B}) + \frac{\rho_E c^2}{4\pi} \nabla^2 \mathbf{B}. \tag{8.1}$$

The first term on the right-hand side describes the convective effect of a conductive fluid dragging and stretching the magnetic lines of force; the second term, the dissipative effect due to the decay of electric current by the resistivity (e.g., Ichimaru, 2004a).

The magnetic Reynolds number R_m in magnetohydrodynamics is a number representing the ratio between the first term and the second. The larger the R_m, the more effectively are the magnetic activities and field strengths sustained and amplified by the fluid motion.

The magnetic Reynolds number associated with this resistivity and Jupiter's self-rotation may then be estimated as

$$R_m = \frac{4\pi R_J^2 \omega_J}{\rho_E c^2} \approx 0.82 \times 10^{12}, \tag{8.2}$$

where ω_J (=15.2 rad/day) represents the angular velocity of Jupiter's self-rotation.

8.1.3 MAGNETIC ACTIVITIES

This value of R_m is to be compared with a corresponding estimate $\sim 1.0 \times 10^8$ for the solar magnetic activities (e.g., Ichimaru, 1996). Such a comparison implies that prominent magnetic activities may be amply sustained near Jupiter by the presence of highly conductive metallic hydrogen inside its MI boundary.

We remark, on the other hand, that there cannot be expected any electrically conductive (i.e., ionized) material in the frigid ($T \approx 129$ K) conditions of the Jovian surface and in its atmosphere, a feature drastically in contrast with those in the solar chromosphere at $T = \sim 6000$ K, where a considerable amount of ionized gas may exist.

The first-order MI transitions in dense hydrogen may thus be looked upon as an element of physics that is essential to the Jovian magnetic activities.

8.2 FERROMAGNETIC AND FREEZING TRANSITIONS IN METALLIC HYDROGEN

In addition to the metal–insulator transitions treated above, another class of phase transitions may be found for hydrogen in metalized states. As with the cases of itinerant electrons or the electron liquids (Ceperley & Alder, 1980; Ichimaru, 2000), the protons in metallic hydrogen may be in a Wigner crystalline state as well as in a paramagnetic or a ferromagnetic fluid state.

We thus consider these issues of ferromagnetic and/or freezing transitions and thereby elucidate the associated phase diagrams for metallic hydrogen (Ichimaru, 2001).

8.2.1 EQUATIONS OF STATE WITH SPIN POLARIZATION

Theoretical approaches to these issues begin with evaluations of the free energies as in (6.11). In the present case, the degrees of ionization and molecular dissociation are to be set at $\langle Z \rangle = 1$ and $\alpha_d = 0$; instead, the spin polarization, $\zeta = (n_\uparrow - n_\downarrow)/n$, with $n_\uparrow$ and $n_\downarrow$ denoting the partial number densities of spin up and spin down protons (where $n = n_\uparrow + n_\downarrow$) enters as a new parameter.

We thus consider the total free energy, $f_{tot}(n, T; \zeta)$, in place of (6.11). The degrees of spin polarization and the resultant magnetic states are determined through minimization of the total free energy with respect to the variation of ζ.

8.2.2 PHASE DIAGRAMS WITH SPIN POLARIZATION

Phase diagrams of hydrogen describing magnetization and solidification of metallic hydrogen are obtained in Figure 8.1 for higher density and finite temperature regime (Ichimaru, 2001).

Table 8.1 lists the values of the physical parameters at the fluid–solid critical point (C_{FS}) and the magnetic critical point (M_C) in the phase diagrams of Figure 8.1.

8.3 NUCLEAR FERROMAGNETISM WITH MAGNETIC WHITE DWARFS

The white dwarf represents a final stage of stellar evolution, illustrated in Figure 1.2. It corresponds to a star of about one solar mass compressed to a characteristic radius (R_{WD}) of approximately 5000 km and an average density of some 10^6–10^7 g/cm^3.

8.3.1 HYDROGEN WITH MAGNETIC WHITE DWARFS

The interior of a white dwarf consists of a multi-ionic condensed matter composed of C and O as the main elements and Ne, Mg, Si, …, Fe as trace elements. Observationally, a class of white dwarfs (DA) possesses hydrogen-rich atmosphere and envelopes.

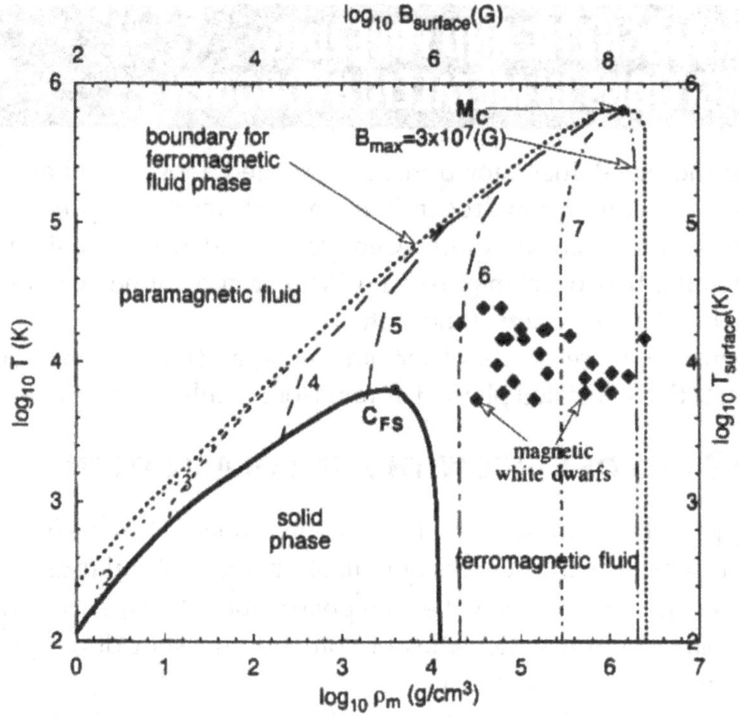

FIGURE 8.1 Phase diagrams of metallic hydrogen describing partial spin ordering and Wigner crystallization. The thick solid curve depicts the phase boundary between fluid and solid, with C_{FS} designating the associated critical point. The dashed and chain curves describe the conditions at a constant strength magnetization, with the numerals denoting the decimal exponents of the field strength B_M in G; M_C designates the magnetic critical point. The diamond markers plot the observed surface-field strengths ($B_{surface}$) vs. the effective surface temperatures ($T_{surface}$) for the magnetic white dwarfs (Ichimaru, 2001).

TABLE 8.1 Physical Parameters at the Fluid–Solid Critical Point (C_{FS}) and the Magnetic Critical Point (M_C) in the Phase Diagrams of Figure 8.1 for the Liquid-Metallic Hydrogen

	C_{FS}	M_C
ρ_m (g/cm³)	4.0×10^3	1.4×10^6
T (K)	6.6×10^3	6.6×10^5
B_M (G)	2.6×10^5	8.0×10^6

Note: Ichimaru 2001

Of all the isolated white dwarfs surveyed, only about 3%–5% have observable magnetic fields that are in the range ~1–500 MG. The surface magnetic fluxes of these white dwarfs, $\sim BR_{WD}^2 = (10^{24}\text{–}5\times10^{26})$ gauss cm², are regarded as similar in magnitude to those of the magnetic Ap stars, a spectroscopic type of stars that exhibit intense hydrogen lines. Such Ap stars may therefore be looked upon as probable progenitors of the magnetic white dwarfs (Chanmugam, 1992; Weisheit, 1995).

8.3.2 ORIGIN OF STRONG MAGNETIZATION

The ferromagnetic transitions in dense metallic hydrogen as elucidated in Figure 8.1 may offer an ingredient of physics relevant to the mechanisms for the origin of strong magnetization in the magnetic white dwarfs (Ichimaru, 1997). The magnetic critical point occurs at a temperature of 6.6×10^5 K, and the strengths of induced magnetization may range as high as 3×10^7 G along the B_{max} line of Figure 8.1.

The plots in Figure 8.1 of the magnetic-field strengths ($B_{surface}$) versus the effective blackbody temperatures ($T_{surface}$) on the surfaces of 25 magnetic white dwarfs observed (Weisheit, 1995) have shown that except for a few cases (with the field strengths exceeding the theoretical maximum by a factor of 1~15) the observed field strengths fall within the predicted range and that the surface temperatures are lower than the critical temperature by approximately two orders of magnitude.

The hydrogen densities (~10^4–10^6 g/cm^3) required for magnetization may be expected in an outer shell of a hydrogen-rich white dwarf at some 70%–90% of the stellar radius, where the mass densities may take on values lower by three or more orders of magnitude than those in the core.

Solidification of metallic hydrogen, on the other hand, may not take place in such a white dwarf, since all the $T_{surface}$ values exceed the temperature at C_{FS}.

Figure 8.2 may possibly provide supporting evidence for hydrogen-rich fluid atmosphere with the magnetic white dwarfs.

8.3.3 FIELD AMPLIFICATION BY STELLAR ROTATION

For those white dwarfs with super-strong magnetic fields exceeding 3×10^7 G, a maximum field strength obtainable by the ferromagnetism alone in metallic hydrogen, a separate amplification mechanism such as a "dynamo" would have to be called for (e.g., Chanmugam, 1992), where the nuclear ferromagnetism in Figure 8.1 may provide "seed" fields for such an amplification.

The electric resistivity ρ_E calculated for the metallic hydrogen at the critical conditions (M_C) takes on the value, 6.0×10^{-12} Ω·cm (Kitamura & Ichimaru, 1996). The magnetic Reynolds number associated with a rotating white dwarf is thus evaluated as

$$R_m = \frac{4\pi R_{WD}^2 \omega_{WD}}{\rho_E c^2} = 3.8 \times 10^{16} \left(\frac{R_{WD}}{5000 \text{ km}} \right)^2 \left(\frac{\omega_{WD}}{2\pi / \text{day}} \right), \tag{8.3}$$

where ω_{WD} is the angular velocity of a white dwarf (cf., Figure 8.2).

Comparing this number with the corresponding estimates, ~1.2×10^{12} and ~1.0×10^8, for Jovian and solar magnetic activities (Ichimaru, 1997) (where the ferromagnetic seed fields are not available) respectively, we may conclude that considerable strengths of

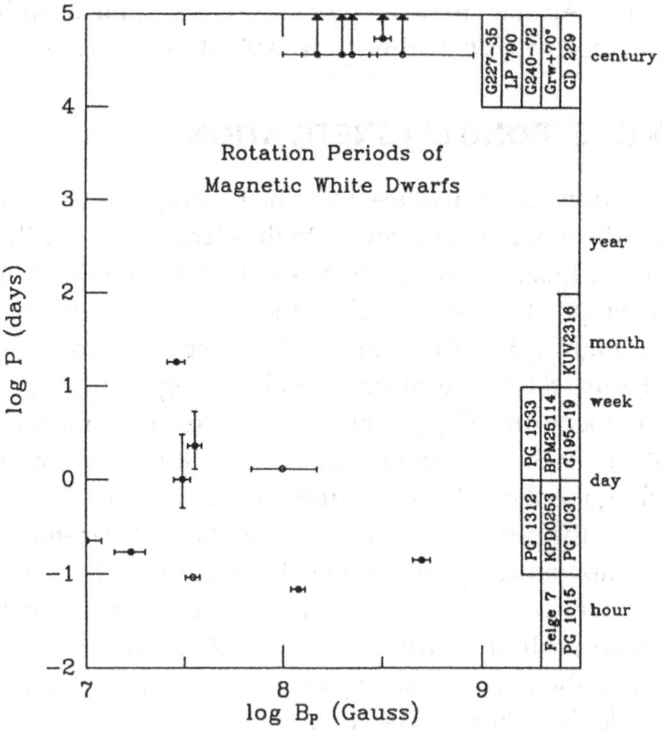

FIGURE 8.2 Rotation period of isolated magnetic white dwarfs as a function of polar magnetic field and as a histogram on right ordinate. An alternative explanation of the five apparently static stars is that they are rotating very rapidly or their fields are axisymmetric. Filled circles correspond to H-rich atmosphere; open circles correspond to mixed or unknown composition (Schmidt & Norsworthy, 1991).

the magnetic fields may be sustained around a white dwarf by the presence of highly conductive, *ferromagnetic* hydrogen in the outer shell.

These observations may possibly account for some of the questions raised by Weisheit (1995) in regard to apparent non-correlation between surface magnetic fields, temperatures, and rotation periods with magnetic white dwarfs.

9

NUCLEAR FUSION IN METALLIC HYDROGEN

Nuclear reactions in hydrogen and its isotopes, deuterium and tritium, are the principal subjects in the development of fusion reactors. Condensed-matter effects including thermodynamics and phase transitions drastically affect the rate of nuclear reactions in ultra-dense hydrogen (e.g., Ichimaru, 1993; Ichimaru & Kitamura, 1999).

In metallic substances under ultrahigh pressures, electrons act to weaken or *screen* Coulomb repulsion between the atomic nuclei. The effects become so conspicuous that rates of nuclear reactions at relatively low temperatures take on values independent of the temperature. Cameron (1959) coined the term "pycnonuclear" reactions (from the Greek, πψκνοσ, meaning "compact, dense") to describe such nuclear processes. These reactions are considered to be applicable to the processes in a white-dwarf progenitor of a supernova, as illustrated in Figure 1.2.

In addition to such screening effects by electrons, the strong spatial correlation between atomic nuclei in dense plasmas due to their Coulomb repulsion, the very cohesive effect manifested in freezing transition, acts to enhance the reaction rates through the effective reduction in overall internuclear repulsion. Such an effect of enhancement in dense fluids increases steeply as the temperature is lowered. The reaction rates in solids, on the other hand, increase sharply with the temperature, as the rates depend sensitively on the amplitude of atomic vibration. Hence, a maximum enhancement may be attained near the conditions of freezing.

In a white-dwarf progenitor of a type Ia supernova, enhancement by a factor of thirty to forty orders of magnitude may thus be anticipated in the rate of nuclear reactions (Salpeter & Van Horn, 1969; Ichimaru, 1993; Ichimaru & Kitamura, 1999). Virtually no significant enhancement is expected, however, in high-temperature, relatively low-density states with the solar interior or with the inertial-confinement fusion (ICF) plasmas.

Construction of the overall phase diagrams describing the metal–insulator (MI) and fluid–solid transitions is therefore pertinent to nuclear fusion in metallic hydrogen under ultrahigh pressure as well; such may possibly lead to a scheme of a "supernova on the Earth" for fusion studies. The phase diagrams in Figure 6.1 may most manifestly illustrate these features of nuclear reactions in dense hydrogen.

9.1 THERMONUCLEAR AND PYCNONUCLEAR REACTIONS

For generality, we consider dense binary-ionic substances with mass density ρ_m, pressure P, and temperature T, consisting of nuclear species with charge number Z_i, mass number A_i, and molar fraction x_i (i = 1, 2). The number density of the nuclei of the species "i" and that of electrons are then given by

$$n_i = \frac{x_i \rho_m}{(x_1 A_1 + x_2 A_2) m_N}, \quad n_e = \frac{(x_1 Z_1 + x_2 Z_2) \rho_m}{(x_1 A_1 + x_2 A_2) m_N}, \tag{9.1}$$

where $m_N = 1.6605 \times 10^{-24}$ g denotes the average mass per nucleon. The ion-sphere radius of (1.6) may here be extended to encompass the binary-ionic systems, so that we define

$$a_{ij} = \frac{1}{2}\left[\left(\frac{3Z_i}{4\pi n_e}\right)^{1/3} + \left(\frac{3Z_j}{4\pi n_e}\right)^{1/3}\right]. \tag{9.2}$$

9.1.1 SCATTERING BY THE COULOMB POTENTIAL

Events of scattering between the nuclei "i" and "j" via the potential, $W_{ij}(r)$, with relative velocity v and reduced mass,

$$\mu_{ij} = \frac{A_i A_j}{A_i + A_j} m_N, \tag{9.3}$$

may be described by the wave functions,

$$\Psi_{ij}(\mathbf{r}) = \sum_{l=0} \Psi_{ij}^{(l)}(r) P_l(\cos\vartheta), \tag{9.4}$$

for the colliding pairs at an inter-nuclear separation **r**; here, $P_l(\cos\vartheta)$ denote the Legendre polynomials (e.g., Landau & Lifshitz, 1965). The wave functions obey the Schrödinger equation,

$$\left[-\frac{\hbar^2}{2\mu_{ij}}\frac{d^2}{dr^2} + W_{ij}(r) + \frac{\hbar^2 l(l+1)}{r^2} - E\right] r\Psi_{ij}^{(l)}(r) = 0, \tag{9.5}$$

where $E = (\mu_{ij}/2)v^2$ denotes the center-of-mass energy.

9.1.2 PROBABILITY OF PENETRATION—BARE COULOMB REPULSION

The penetration probability $p(E)$ of the colliding nuclei to a nuclear reaction radius r_N, proportional to $|\Psi_{ij}(r_N)|^2$, may be obtained by the solution to the Schrödinger equation. With the bare Coulomb potential,

$$W_{ij}(r) = \frac{Z_i Z_j e^2}{r}, \tag{9.6}$$

substituted in (9.5), the essential parameters characterizing Coulomb scattering in the short ranges are the *nuclear Bohr radius*,

$$r_{ij}^* = \frac{\hbar^2}{2\mu_{ij} Z_i Z_j e^2}, \tag{9.7}$$

and the *Gamow energy*,

$$E_G = \frac{Z_i Z_j e^2}{r_{ij}^*} \approx 49.5\left(Z_i Z_j\right)^2 \frac{2\mu_{ij}}{m_N} \quad (\text{keV}). \tag{9.8}$$

The events of scattering are governed primarily by the effective potential between the nuclei in the short-range domain, where the potential may be regarded as isotropic and Coulombic, irrespective of the inter-particle configurations. Calculation of the reaction rates may be facilitated by the observation that the major contributions to the penetration probabilities arise from the *s*-wave ($l=0$) scattering between the reacting nuclei, as wave functions in a spherically symmetric potential with azimuthal quantum number l are proportional to r^l in the short ranges.

Since one can generally assume that the nuclear reaction radius $r_N < r_{ij}^*$, the *s*-wave scattering is the major contribution to the reaction rates; hence, $r_N \approx 0$ may be taken for the calculation of the penetration probabilities, $p(E)$. In these connections, one notes the short-range cusp condition with Coulomb scattering,

$$\lim_{r\to 0} \frac{d \ln \Psi_{ij}(r)}{dr} = \frac{1}{2r_{ij}^*}. \tag{9.9}$$

The usual boundary conditions in a treatment of these scattering problems assume an incident plane wave in the z direction. The asymptotic ($r \to \infty$) form of the Coulomb wave function is then expressed as (e.g., Landau & Lifshitz, 1965)

$$\Psi_{ij}(\mathbf{r}) \to \exp\left[i\kappa z + i\eta \ln \kappa(r-z)\right]\left[1 + \frac{\eta^2}{i\kappa(r-z)}\right]$$

$$+ \frac{f_c(\vartheta)}{r} \exp\left[i(\kappa z - \eta \ln 2\kappa r)\right]$$

where

$$f_c(\vartheta) = \frac{\Gamma(1+i\eta)}{i\Gamma(-i\eta)} \frac{\exp\left[-i\eta\ln\left(\sin^2\frac{1}{2}\vartheta\right)\right]}{2\kappa\sin^2\frac{1}{2}\vartheta}$$

is the angular function of the scattering wave,

$$\kappa \equiv \frac{\mu_{ij}v}{\hbar}, \quad \eta \equiv \frac{Z_iZ_je^2}{\hbar v}, \tag{9.10}$$

the radial coordinate r with respect to the scattering center represents the distance between the reacting nuclei, and $\Gamma(z)$ refers to the gamma function defined by (AIV.8).

With a normalization such that

$$\int_\Omega d\mathbf{r}\left|\Psi_{ij}(\mathbf{r})\right|^2 = \Omega$$

over a spherical volume Ω with a radius far greater than $2a_{ij}$, the incident flux at $z \to -\infty$ is given by

$$\frac{\hbar}{2i\mu_{ij}}\left[\Psi_{ij}^*\left(\nabla\Psi_{ij}\right) - \Psi_{ij}\left(\nabla\Psi_{ij}\right)^*\right] = v. \tag{9.11}$$

The penetration probability or the square of the wave function at the origin then takes on the values

$$\begin{aligned} p(E) = \left|\Psi_{ij}(0)\right|^2 &= \frac{1}{v}\left|\Gamma(1+i\eta)\right|^2\exp(-\pi\eta) \\ &= \frac{2\pi\eta}{\exp(2\pi\eta)-1} = \frac{\pi\sqrt{E_G/E}}{\exp\left(\pi\sqrt{E_G/E}\right)-1}. \end{aligned} \tag{9.12}$$

This function approaches unity in the high-energy limit, $E \gg E_G$, as it should.

9.1.3 CROSS-SECTION FACTOR

In the low-energy ($E \ll E_G$) processes of astrophysical interest, the penetration probability of the Coulomb barrier vanishes exponentially as

$$p(E) \to \pi\sqrt{\frac{E_G}{E}}\exp\left(-\pi\sqrt{\frac{E_G}{E}}\right).$$

Hence, one singles out this exponential factor, representing a purely Coulombic effect, from the cross-section $\sigma_{ij}(E)$ for the nuclear reactions between species i and j, and thereby introduces the *cross-section factor*, $S_{ij}(E)$, via (Salpeter, 1952; Barnes, 1971; Fowler, 1984)

$$\sigma_{ij}(E) = \frac{S_{ij}(E)}{E} \exp\left(-\pi\sqrt{\frac{E_G}{E}}\right). \tag{9.13}$$

Cross-section factors therefore quantify the proficiency of reactions *intrinsic* to the nuclei.

For nonresonant nuclear reactions, $S_{ij}(E)$ are functions slowly varying with E and may be expressed as

$$S_{ij}(E) = S(0)\left\{1 + \frac{S'(0)}{S(0)}E + \frac{1}{2}\frac{S''(0)}{S(0)}E^2\right\}.$$

For isotopes of hydrogen, we list in Table 9.1 the values of the cross-section factors and the energies Q_{ij} released per reaction for the reactions:

$$d + d \to t + p,$$

$$d + d \to {}^3\mathrm{He} + n,$$

$$t + d \to {}^4\mathrm{He} + n.$$

9.1.4 PROBABILITY OF PENETRATION—SCREENED COULOMB REPULSION

In a dense metallic system, electrons act to screen the Coulomb repulsion between atomic nuclei. The potential $W_{ij}(\mathrm{r})$ of scattering in (9.5) now deviates from the purely Coulombic form (9.6) and takes on values,

$$W_{ij}(r) = \frac{Z_i Z_j e^2}{2\pi^2}\int d\mathbf{k}\,\frac{\exp(-i\mathbf{k}\cdot\mathbf{r})}{k^2\varepsilon_e(k,0)} \to \frac{Z_i Z_j e^2}{r}\left[1 - \frac{r}{D_s}\cdots\right]. \tag{9.14}$$

For non-relativistic electrons ($r_s \geq 0.1$), we employ the local-field corrections in $\varepsilon_e(k,0)$ (Ichimaru & Utsumi, 1981, 1983), accounting for the strong Coulomb-coupling effects

TABLE 9.1 Nuclear Reaction Cross-Section Factors and Q_{ij} Values for Isotopes of Hydrogen

Reactions	$S(0)$ (MeV·barn)	$S'(0)/S(0)$ (MeV^{-1})	$S''(0)/S(0)$ (MeV^{-2})	Q_{ij} (MeV)
d(*d*,*p*)*t*	0.0530	4.95	–	4.033
d(*d*,*n*)^{3}He	0.0530	4.95	–	3.269
t(*d*,*n*)^{4}He	11.0	13.8	623	17.590

Note: 1 barn = 10^{-24} cm^2

of Sec. 2.2.7 between electrons, and evaluate the short-range screening distance D_s of the electrons, defined by (6.13), as

$$\frac{a}{D_s} = 1.239 r_s^{v} \left(\tanh \frac{1.061}{\theta} \right)^{1/2} \tag{9.15}$$

with

$$\nu = \frac{0.435 + 0.024\theta^{2.65}}{1 + 0.048\theta^{2.65}}.$$

For relativistic electrons ($r_s < 0.1$ and $\theta << 0.1$), the screening parameter in (6.14) has been computed (Ichimaru & Utsumi, 1981, 1983) with Jancovici dielectric function (Jancovici, 1962) as

$$\begin{aligned}\frac{a}{D_s} = {} & 0.1718 + 0.9283 r_s + 1.591 \times 10^2 r_s^2 - 3.800 \times 10^3 r_s^3 \\ & + 3.706 \times 10^4 r_s^4 - 1.307 \times 10^5 r_s^5, \quad (r_s < 0.1).\end{aligned} \tag{9.16}$$

It is noteworthy that the screening length (in units of a) takes on the finite value 5.8 in the limit of high densities ($r_s \to 0$), while (9.15) and the nonrelativistic Thomas–Fermi length as obtained from (2.45) diverge in the same limit. Due to such relativistic effects, the electron screening may thus remain considerable in dense stellar materials.

Substituting (9.14), rather than (9.6), in (9.5), we find that the *pycnonuclear penetration probability* is now given by

$$p(E) = \frac{\pi \sqrt{E_G / (E + E_s)}}{\exp\left(\pi \sqrt{E_G / (E + E_s)} \right) - 1}. \tag{9.17}$$

Here,

$$E_s = \frac{Z_i Z_j e^2}{D_s} \tag{9.18}$$

is the Coulomb energy associated with the screening distance. It should be remarked here that $p(E)$ in (9.17) takes on a non-vanishing value,

$$p(0) = \frac{\pi \sqrt{E_G / E_s}}{\exp\left(\pi \sqrt{E_G / E_s} \right) - 1}, \tag{9.19}$$

in the limit, $E \to 0$.

This is a feature unique in the pycnonuclear reactions, markedly different from the thermonuclear case of (9.12); the limiting value (9.19) decreases exponentially as the reduced mass increases. It offers the very reason why the proton-deuteron (p-d) pycnonuclear reactions should be pursued in metallic hydrogen.

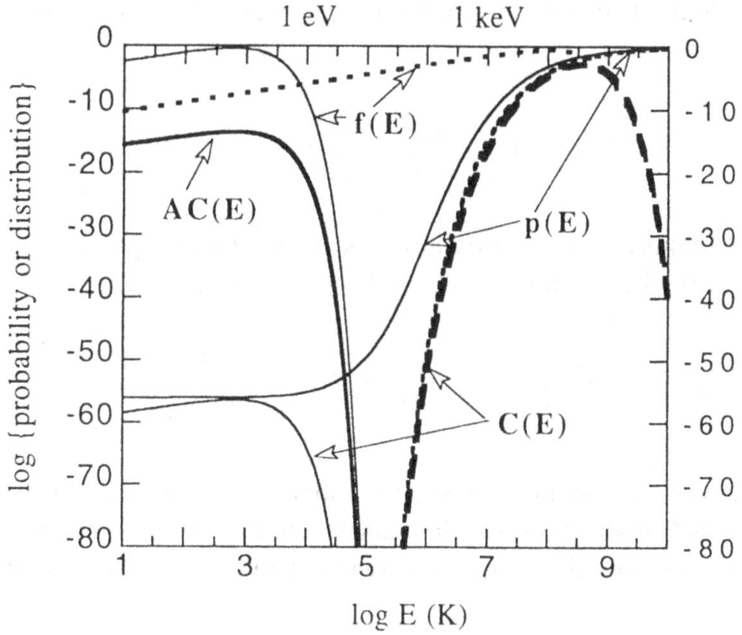

FIGURE 9.1 Dashed and dotted curves: thermonuclear (*d-t*, at 10^8 K) vs. solid curves: pycnonuclear (*p-d*, at 450 K with $n_e = 2.4 \times 10^{24}$ cm^{-3}) reactions; p(E) denotes the penetration probability through the Coulomb barrier between a pair of reacting nuclei with a center-of-mass energy, E(K), in temperature units; f(E), the Boltzmann distribution; C(E) = p(E)·f(E), a contact probability between a pair of reacting nuclei; A·C(E), enhanced contact probability with A (= 5.7×10^{42}) denoting the enhancement factor (9.32).

In Figure 9.1, we plot and compare the penetration probabilities, the Boltzmann distributions,

$$f(E) = 2\pi \left(\frac{E}{\pi T} \right)^{3/2} \exp\left(-\frac{E}{T} \right),$$

and the contact probabilities, $C(E) = p(E) \cdot f(E)$, as functions of log E(K), the logarithmic energy in temperature units, for the deuterium-tritium (*d-t*) thermonuclear reactions at $T = 10^8$ K and for the *p-d* pycnonuclear reactions at $T = 450$ K with the electron density, $n_e = 2.4 \times 10^{24}$ cm^3.

9.1.5 RATES OF THERMONUCLEAR REACTIONS

The rates of thermonuclear reactions stemming from those scattering processes may be calculated by the thermal averages $\langle \sigma_{ij}(E) v \rangle$ of (9.13) over the Boltzmann distribution,

$$f_B(E) = \frac{2}{k_B T} \left(\frac{E}{\pi k_B T} \right)^{1/2} \exp\left(-\frac{E}{k_B T} \right). \quad (9.20)$$

The result yields the Gamow rates (Gamow & Teller, 1938; Thompson, 1957) of the thermonuclear reactions,

$$P_G(\rho_m, T) = \frac{16 Q_{ij} S_{ij} r_{ij}^* \tau_{ij}^2}{3^{5/2}\pi(1+\delta_{ij})\hbar} n_i n_j \exp(-\tau_{ij}). \tag{9.21}$$

Here, S_{ij} refers to a thermal average of the cross-section factor, δ_{ij} denotes the Kronecker delta distinguishing between the cases with $i \neq j$ and with $i = j$;

$$\tau_{ij} = 3\left(\frac{\pi^2 E_G}{4k_B T}\right)^{1/3} \approx 33.70\left(Z_i Z_j\right)^{2/3}\left(\frac{2\mu_{ij}}{m_N}\right)^{1/3}\left(\frac{T}{10^6\ \mathrm{K}}\right)^{-1/3} \tag{9.22}$$

refers to the Gamow exponent. The reaction rate (9.21) contains a factor $\exp(-\tau_{ij})$. The magnitude of this Gamow exponent increases with the charge numbers and/or with the reduced masses. The thermonuclear reaction rates vanish exponentially as the temperature decreases.

The integration leading to the Gamow rate contains in its integrand a product between a steeply rising cross-section $\sigma_{ij}(E)$ and a steeply decreasing Boltzmann distribution $f_B(E)$ as functions of E. The product thus exhibits a *Gamow peak* at the energy,

$$E_{GP} = \frac{1}{3}\tau_{ij} k_B T. \tag{9.23}$$

The radius r_{TP} of the classical turning point for a colliding pair with the Gamow peak energy is given by

$$r_{TP} = \frac{3 Z_i Z_j e^2}{\tau_{ij} k_B T}. \tag{9.24}$$

The S_{ij} in (9.21) is thus to be evaluated as $S_{ij}(E_{GP})$. In Figure 9.1, we observe a Gamow peak at $E \approx 3 \times 10^8$ K for the contact probability in the *d-t* thermonuclear reactions.

9.1.6 RATES OF PYCNONUCLEAR REACTIONS

For the pycnonuclear reactions, where the penetration probability is now given by (9.19), the screening temperature, derived from setting $E_s = E_{GP}$, may be expressed as (Ichimaru, 1993; Ichimaru & Kitamura, 1999)

$$T_s\ (\mathrm{K}) = 5.7 \times 10^4\ (\mathrm{K})\sqrt{\frac{Z_i Z_j m_N}{2\mu_{ij}}}\left(\frac{D_s}{10^{-9}\ \mathrm{cm}}\right)^{-3/2}. \tag{9.25}$$

In the weak screening regime such that $T > T_s$, the rate of reactions is calculated through a perturbative modification of Gamow's thermonuclear rate. Thus, the enhancement factor due to such weak screening is

$$A_{ij}^{(e)}(\rho_m, T) = \left(1 - \frac{3E_s}{\tau_{ij} k_B T}\right) \exp\left(\frac{E_s}{k_B T}\right). \tag{9.26}$$

The rate of thermonuclear reactions is then calculated as the product between the Gamow rate (9.21) and the enhancement (9.26).

In the strong screening regime $T < T_s$, the pycnonuclear counterpart to this product is given by (Ichimaru, 1993; Ichimaru & Kitamura, 1999)

$$P_s(\rho_m, T) = \frac{2Q_{ij} S_{ij} r_{ij}^*}{(1 + \delta_{ij})\hbar} n_i n_j \sqrt{\frac{D_s}{r_{ij}^*}} \exp\left(-\pi \sqrt{\frac{D_s}{r_{ij}^*}}\right). \tag{9.27}$$

In this case, the classical turning point is located at $r_{TP} = D_s$. The screening distance, given by (9.15) or (9.16), is virtually independent of T in a dense material, because $\theta < 0.1$.

Contrary to the Gamow rate (9.21), which changes sharply with the temperature through τ_{ij}, the pycnonuclear rate (9.27) is practically *independent* of the temperature; the strong exponential decrease with D_s, the charge product $Z_i Z_j$, and the reduced mass should be noted.

9.2 SOLAR PROCESSES AND INERTIAL-CONFINEMENT FUSION

In the Sun, nuclear reactions take place most vigorously near the core, where the mass density and the temperature of the metallic hydrogen are estimated to be 56.2 g/cm^3 and 1.55×10^7 K (see Figure 6.1).

9.2.1 INERTIAL-CONFINEMENT FUSION

Terrestrial inertial-confinement fusion researchers are presently attempting to realize conditions analogous to those near the solar core by the compression of *d-t* fuel to mass densities and temperatures on the order of 3–60 g/cm^3 and ~10^8 K; hence, a "sun on the Earth." At such a high temperature, required for the thermonuclear reactions, one anticipates an assortment of dynamic instabilities to overcome during the compression process of hydrogen matter.

9.2.2 THE *P–P* CHAIN

One of the proton–proton chains, the fundamental nuclear processes in the solar interiors, consists of

$$p(p, e^+ \nu_e) d(p, \gamma)^3\text{He}\left(^3\text{He}, 2p\right)^4\text{He},$$

which altogether yields

$$4p \rightarrow \alpha + 2e^+ + 2\upsilon_e + 26.2(\text{MeV}).$$

The cross-section factors S_{ij} and the Q_{ij} values are (Bahcall & Ulrich, 1988)

$$S_{pp} = 4.07 \times 10^{-25}\,(\text{MeV·barn}), \quad Q_{pp} = 1.442\,(\text{MeV}),$$

$$S_{pd} = 2.5 \times 10^{-7}\,(\text{MeV·barn}), \quad Q_{pd} = 5.494\,(\text{MeV}),$$

$$S_{^3\text{He}^3\text{He}} = 5.15 \times 10^{0}\,(\text{MeV·barn}), \quad Q_{^3\text{He}^3\text{He}} = 12.860\,(\text{MeV}).$$

The *p–p* chain, starting with $p(p,e^+\nu_e)d$, involves a β process and thus is extremely slow; the rate of this chain is controlled by these slow processes.

9.3 ENHANCEMENT OF NUCLEAR REACTIONS IN METALLIC FLUIDS

As a progenitor of the type Ia supernova, a white dwarf with interiors consisting of a carbon–oxygen mixture can be considered a kind of binary-ionic mixture, with a central mass density of 10^7 to 10^{10} g/cm^3 and a temperature of 10^7 to several times 10^9 K (Starrfield et al., 1972; Whelen & Iben, 1973). Nuclear runaway leading to supernova explosion may take place when the thermal output due to nuclear reactions exceeds the rate of energy losses.

9.3.1 ENHANCEMENT DUE TO COULOMB CORRELATION

In addition to the electronic screening effects manifested in the pycnonuclear penetration probability (9.17), strong Coulomb correlation between atomic nuclei in dense matter acts to enhance the reaction rates through an effective reduction of the internuclear repulsion (e.g., Ichimaru, 1993; Ichimaru & Kitamura, 1999). It is the very correlation effect responsible for the freezing transitions considered earlier in Secs. 5.2, 5.4, and 6.3, and it is closely related to the Coulombic chemical potentials in dense plasmas treated in Secs. 5.2 and 5.3.

The increment of the Coulombic free energy before and after the reaction is expressed as

$$\Delta F_{ij} = \mu_{\text{Coul}}(Z_i + Z_j) - \mu_{\text{Coul}}(Z_i) - \mu_{\text{Coul}}(Z_j), \tag{9.28}$$

where $\mu_{\text{Coul}}(Z_i)$ denotes the Coulombic chemical potential of a charge Z_i. The increment before and after the nuclear fusion ΔF_{ij} is a *negative* quantity, expressing a cohesive effect in Coulombic matter; its magnitude increases proportionally to the cubic root of the matter density.

It is instructive at this stage to recall the *ion-sphere model* (e.g., Ichimaru, 1982) as illustrated in Figure 1.5. We construct an ion sphere by picking a particle (an ion with the electric charge Ze) in the plasma and by associating with it a sphere of neutralizing charges that would exactly cancel the point charge of the ion. This sphere has the radius a and the electric charge density $-3Ze/4\pi a^3$. The electrostatic energy of the ion sphere is then calculated to be $-0.9(Ze)^2/a$; the increment before and after the nuclear fusion is thus $-1.057(Ze)^2/a$ in the ion-sphere model (Salpeter & Van Horn, 1969).

9.3.2 ENHANCEMENT FACTOR

The enhancement factor for the rate of nuclear reactions in dense metallic fluid is expressed approximately as

$$A_{ij} \approx \exp\left(-\frac{\Delta F_{ij}}{k_B T}\right).$$

In the ion-sphere model, it takes on a huge value, $\exp(1.057\Gamma)$, for a strongly coupled plasma, as Figure 9.2 explains.

The quantity ΔF_{ij} may be calculated more accurately in terms of the ion–ion correlation energies as

$$\frac{\Delta F_{ij}}{k_B T} = f_i\left(\Gamma_{i+j}, \Theta_{ij}\right) - f_i\left(\Gamma_i, \Theta_{ij}\right) - f_i\left(\Gamma_j, \Theta_{ij}\right), \quad (9.29)$$

where the so-called "linear mixing law" for the interaction energies, applicable fairly accurately to dense plasmas (e.g., Ichimaru, 1993), has been adopted. In (9.29), we

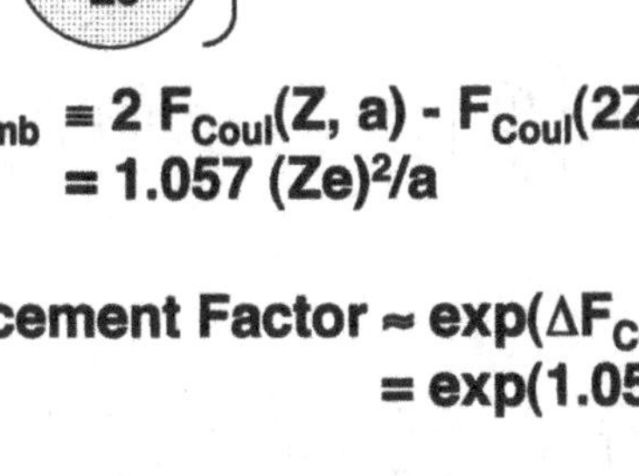

$$\Delta F_{Coulomb} = 2\, F_{Coul}(Z, a) - F_{Coul}(2Z, 2^{1/3} a) = 1.057\,(Ze)^2/a$$

$$\text{Enhancement Factor} \sim \exp(\Delta F_{Coul}/k_B T) = \exp(1.057\,\Gamma)$$

FIGURE 9.2 Origin of the cohesive force and the enhancement factor in a dense plasma.

employ a semiclassical expression (5.11) for the ion–ion correlation energy applicable for $\Gamma >> 1$ with spin-independent, quantum-statistical corrections (Ichimaru, 1997),

$$f_i(\Gamma,\Theta) = -0.895929\Gamma + \left(0.0678763\frac{\Gamma}{\Theta} - 0.000621921\frac{\Gamma^2}{\Theta^2}\right)\exp\left(-\frac{16}{\Theta^2}\right)$$
$$+0.402187\Gamma^{1/2} + 1.193208 - 3.426161\Gamma^{-1/2}, \tag{9.30}$$

where

$$\Theta_{ij} \equiv \frac{4\mu_{ij}k_BT}{\hbar^2\left[3\pi^2(n_i+n_j)\right]^{2/3}};$$

Γ_i is the electron-screened Coulomb coupling parameter for the nuclei with charge Z_i, that is,

$$\Gamma_i \equiv \frac{Z_i^{5/3}e^2}{a_e k_B T}\exp\left(-0.85Z_i^{1/3}\frac{a_e}{D_s}\right) \tag{9.31a}$$

with $a_e = (3/4\pi n_e)^{1/3}$ and

$$\Gamma_{i+j} \equiv \frac{\left(Z_i+Z_j\right)^{5/3}e^2}{a_e k_B T}\exp\left[-0.85\left(Z_i+Z_j\right)^{1/3}\frac{a_e}{D_s}\right]. \tag{9.31b}$$

As we have noted in Sec. 9.1, the rate of nuclear reactions are proportional to the statistical averages of penetration or contact probabilities, $|\psi_{ij}(0)|^2$, which are in fact the joint probability densities, $g_{ij}(r)$, between nuclei "i" and "j" evaluated at a distance of nuclear force, r_N (≈ 0). Enhancement factors for thermonuclear or pycnonuclear rates have been calculated through the quantum statistical treatments of such joint probability densities (Alastuey & Jancovici, 1978; Ogata, Iyetomi, & Ichimaru, 1991; Ichimaru, 1993; Ichimaru & Kitamura, 1999); the results are expressed compactly as

$$A_{ij}(\rho_m, T) = \exp(\xi_{ij}) \tag{9.32}$$

where

$$\xi_{ij} = -\frac{\Delta F_{ij}}{k_B T} - \frac{5}{32}\Gamma_{ij}\left(\frac{r_{TP}}{a_{ij}}\right)^2\left[1 + \left(C_1 + C_2\ln\Gamma_{ij}\right)\frac{r_{TP}}{a_{ij}} + C_3\left(\frac{r_{TP}}{a_{ij}}\right)^2\right] \tag{9.33}$$

with

$$\Gamma_{ij} = \frac{Z_i Z_j e^2}{a_{ij}k_B T}\exp\left(-0.85\frac{a_{ij}}{D_s}\right), \tag{9.34}$$

TABLE 9.2 Rates of Nuclear Reactions and Enhancement

Case	Solar Core	ICF	White Dwarf	Liquid Metal 1	Liquid Metal 2
Reactions	*p-p* chain	*d-t*	^{12}C-^{12}C	*p-d*	*p-d*
ρ_m (g/cm^3)	56.2	5	4×10^9	6	9
T (K)	1.55×10^7	10^8	5×10^7	450	500
Γ_{ij}	0.040	0.003	87	297	324
log A_{ij}	0.005	0.000	36.02	42.76	50.27
log P_{ij} (W/g)	−6.48	19.07	−15.97	0.72	10.54

Note: For the Calculations of ^{12}C-^{12}C Reactions in a White Dwarf, the Enhancement Factor Is Given by the Product between (9.26) and (9.31); $S_{C\text{-}C}=8.83\times10^{16}$ (MeV·bahn) and $Q_{C\text{-}C}=13.931$ (MeV) Have Been Assumed

$C_1=1.1858$, $C_2=-0.2472$, and $C_3=-0.07009$ (Ichimaru, 1991). Note that the enhancement factor (9.32) depends sensitively on ρ_m and T; it increases with ρ_m, and sharply as T decreases.

9.3.3 RATES OF NUCLEAR REACTIONS IN DENSE PLASMAS

In Table 9.2, we list the rate of nuclear reactions P_{ij} (expressed in power per unit mass) and the enhancement factor A_{ij} computed for a hydrogen plasma appropriate to the solar core, an ICF deuteron-triton plasma with equal molar fractions, dense ^{12}C matter expected in a white dwarf, and cases of metallic hydrogen with equal molar fractions of protons and deuterons. We observe huge enhancement factors for the "white dwarf" and "liquid metal" cases. As we find in Figure 9.1, the net contact probability, $A\cdot C(E)$, for the "liquid metal" case may assume a magnitude comparable to that for the "ICF" case.

9.4 "SUPERNOVA ON THE EARTH"

The possibilities of combined utilization of the pycnonuclear *p-d* reactions at lower temperatures and their enhancement due to the strong Coulomb correlation, both applicable in ultradense metallic hydrogen near freezing conditions, have led to a proposal of a "supernova on the Earth" scheme for nuclear fusion researches (Ichimaru, 1991, 1993; Ichimaru & Kitamura, 1998, 1999). The idea is to bring a *p-d* mixture to a liquid-metallic state near solidification, as illustrated in Figure 6.1.

9.4.1 ADIABATIC COMPRESSION

For a concrete example, let us consider an experimental scheme of the compression and metallization of a *p-d* mixture with equal molar fractions to a final state: $\rho_m\approx6$ g/cm^3, $T\approx459$ K, $P\approx50$ Mbar (the case "liquid metal 1" in Table 9.2).

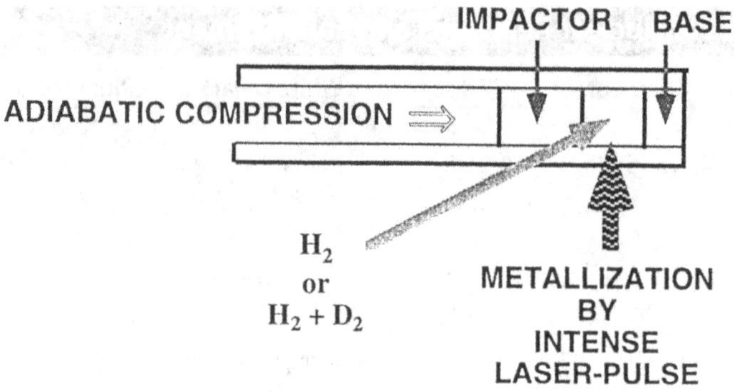

FIGURE 9.3 A schematic diagram of compression/metallization experiments for hydrogen.

As illustrated in Figure 9.1, a considerable rate of enhanced pycnonuclear reactions may be expected in such a *p-d* mixture.

To achieve such an end, we may start from a H_2–D_2 mixture in a low-entropy, molecular-solid (insulator) state at ~1 bar and ~10 K. As shown in Figure 7.4, we find the initial state here is connected by an adiabat to a state with $P \approx 2.4$ Mbar, $T \approx 160$ K, $\rho_m \approx 0.84$ g/cm^3, on the insulator side of the MI coexistence curves. A reverberating shock imparted by a low-speed (~1 μm/ns) impactor with a kinetic energy ~20 kJ, as depicted schematically in Figure 9.3, may thus be utilized for the compression of such a H_2–D_2 mixture with a total mass of ~4.8 mg in a volume of 32 mm^3 (=2 mm × 16 mm^2), say.

The compression may bring the mixture to a state at ~2.4 Mbar and ~220 K on the insulator side of the MI coexistence curves; the enthalpy increment ΔW in such a compression process may amount to ~0.9 kJ. Since the impactor speed is lower, we expect a departure from the adiabat to an extent lesser than those experienced in the Livermore shock-compression experiments (Weir, Mitchel, & Nellis, 1996; Da Silva et al., 1997); thus, the shock compression proposed here may be looked upon as a technical extension in line with these experiments.

9.4.2 METALLIZATION

The ingredient indispensable to such a compression/metallization scheme may then be an injection of a super-intense, ultrashort laser pulse into the compressed hydrogen, at the instant of compression, to ensure an efficient metallization. We estimate the enthalpy ΔW necessary for the metallization is approximately 11 J.

Since the laser pulse-width must be significantly shorter than the time for metallization τ_E (~10 ps), the required laser power should exceed ~1.1 TW. In this regard, the scheme may still be looked upon within the range of technical feasibility.

9.4.3 FEASIBILITY EXPERIMENT

The quasi-adiabatic compression exerted continuously by the slow impactor may bring the resultant liquid-metallic hydrogen finally into a further compressed state at ~50 Mbar and ~450 K, say, near the freezing conditions; ΔW in this compression would be ~3.0 kJ. Contrary to the ultrahigh-temperature ICF plasmas, the dense hydrogen in a liquid-metallic state near solidification is a stable object; no dynamic instabilities are expected.

In this final state, the estimated fusion power would be ~5.2 W/g (Table 9.2), a detectable level for the rate of nuclear reactions. Recent experimental and theoretical progress in ultrahigh-pressure metal physics may make such a scheme of detecting the astrophysical enhanced pycnonuclear reactions, for the first time in the terrestrial laboratory, an attractive and realizable project for fusion studies (Ichimaru & Kitamura, 1995, 1998).

9.4.4 POWER-PRODUCTION EXPERIMENT

With such a feasibility experiment successfully conducted, a further extension into a parameter regime of still higher pressure and density would eventually lead to a fusion scheme with net power production.

For example, if compression to the case "liquid metal 2" (at a pressure of 112 Mbar, still far below an ICF pressure) in Table 9.2 is realized with an increase of the impactor mass and energy, we might expect a burst of power production at a rate ~34.7 GW/g. The actual duration for such a nuclear burning would depend sensitively on the thermal evolution of the hydrogen fuel (Kitamura & Ichimaru, 1996). Assuming that the reactions might last for ~1 ms at this rate, we would find a total thermal output of ~0.17 MJ, which would correspond to a burning ratio of $\sim 2 \times 10^{-4}$.

10

PHASE DIAGRAMS OF NUCLEAR MATTER

Ordinary nuclear matter, when heated or compressed sufficiently, is expected to turn into a new state—a *quark–gluon plasma*—in which the fundamental degrees of freedom are the quarks that compose neutrons and protons, and, at finite temperature, antiquarks and gluons as well. In a sense, the physical circumstances involved are analogous to the metal–insulator transition as described in Sec. 6.5. We here present our current understanding of the phase transition from ordinary matter to the quark–gluon plasma. We then describe the physical situations, such as the early universe, the core of a neutron star, and ultra-relativistic heavy-ion collisions, where such plasmas may be expected.

10.1 DECONFINEMENT OF QUARKS FROM NUCLEONS

The quark–gluon plasma is a phase of matter whose elementary constituents are the quarks, antiquarks, and gluons that make up the strongly interacting particles. It is a new phase in the sense that it has not yet been detected in the laboratory.

10.1.1 RELATIVISTIC HEAVY ION COLLIDER EXPERIMENTS

To probe the densest states of nuclear matter, the nuclear physics community has embarked on a large-scale program of studying collisions of ultra-relativistic heavy ions (e.g., Stenlund et al., 1994; Baym, 1995; Yagi, Hatsuda, & Miake, 2005). In the program, the Relativistic Heavy Ion Collider (RHIC) at Brookhaven National Laboratory plans to provide the capacity of colliding nuclei as heavy as Au on Au at 100 GeV per nucleon in the center of mass (equivalent to 20 TeV per nucleon in a fixed-target experiment).

Recently, the STAR Collaboration (2017) at RHIC reported the first measurement of the rotation of the quark–gluon plasma produced in heavy-ion collisions and thereby provided crucial information for theoretical models that attempt to account for how such plasmas may be formed.

10.1.2 THE OLDEST PHASE OF MATTER

The quark–gluon plasma is said to be the oldest phase of matter, the form of the matter that filled up the early universe at temperatures of trillions of degrees, until the first few microseconds after the Big Bang.

It is also said that the quark–gluon plasma may be found deep in the cores of neutron stars, where an extremely high density of matter is expected.

10.2 PHASES OF NUCLEAR MATTER

We find it instructive to draw an analogy between metallization of hydrogen atoms, treated in Sec. 6.5, and *deconfinement* of quarks from nucleons, as we construct the phase diagrams for nuclear matter, relative to the phase diagrams of hydrogen (Figure 6.1). For example, a proton, a kind of nucleon, is said to assume a confined state of three quarks of different colors.

10.2.1 PHASE DIAGRAMS

Earlier in 1995, G. Baym advanced phase diagrams of nuclear matter, noticing the correspondence between hadrons and insulators on the one hand and between quark–gluon plasmas and metals on the other. He additionally remarked on the gas–liquid phase transitions in hadrons and discontinuous deconfinement transitions in nuclear matter (Baym, 1995).

Recently, Yagi, Hatsuda, and Miake (2005) advanced schematic phase diagrams of nuclear matter, shown in Figure 10.1, in line with Baym's predictions.

Here, "Hadron," "QGP" and "CSC" denote the hadronic phase, the quark–gluon plasma, and the color-superconducting phase, respectively; ρ_{nm} denotes the baryon density of the normal nuclear matter. Possible locations at which we may find the various phases of nuclear matter include hot plasmas in the early universe, dense plasmas in the interiors of neutron stars, and the hot/dense matter created in heavy ion collisions (HICs).

10.2.2 DECONFINEMENT VERSUS METALLIZATION

In Figure 10.1, we also note the critical point associated with the discontinuous deconfinement transitions, analogous to the critical point C_{MI} in Figure 6.1 associated with discontinuous metal–insulator transitions in hydrogen (Kitamura & Ichimaru, 1998).

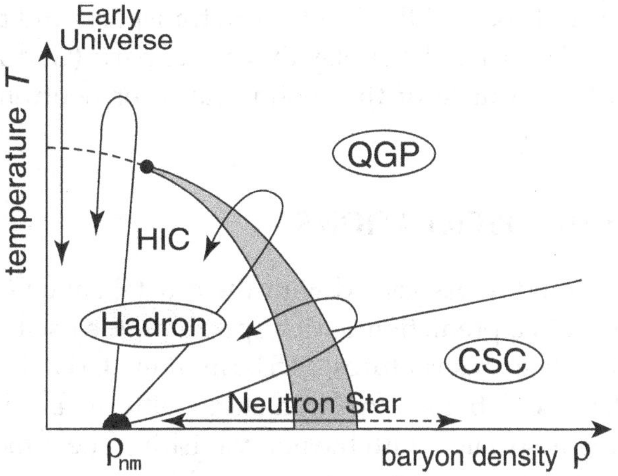

FIGURE 10.1 Schematic phase diagram of nuclear matter. "Hadron," "QGP," and "CSC" denote the hadronic phase, the quark–gluon plasma, and the color-superconducting phase, respectively; ρ_{nm} denotes the baryon density of the normal nuclear matter. Possible locations at which we may find the various phases of nuclear matter include the hot plasma in the early universe, dense plasma in the interior of neutron stars, and the hot/dense matter created in heavy ion collisions (HICs) (Yagi, Hatsuda, & Miake, 2005).

10.3 STRUCTURE OF A NEUTRON STAR

A neutron star is an astronomical object of about one solar mass compressed to a radius of approximately 10 km with an average density well in excess of 0.1 billion tons/cm^3, comparable to the mass density of atomic nuclei ρ_{nm}.

10.3.1 THREE-PART STRUCTURE

The neutron star, like the planet Jupiter in Figure 1.1, may have a three-part structure, consisting of the ultra-dense central core of quark–gluon plasmas, the main body of condensed neutron liquids, and the outer-layer of Wigner-crystallized Coulomb solids consisting mostly in Fe nuclei.

We depicted in Figure 1.3 a schematic structure of a neutron star. According to model calculations, it has an *outer crust*, consisting mostly of iron, with a thickness of several hundred meters and a mass density in the range of $10^4 \sim 10^7$ g/cm^3. At these densities, iron atoms are completely ionized, so each contributes 26 conduction electrons to the system. At temperatures near 10^7 K, the thermal de Broglie wavelengths of the resultant Fe nuclei are substantially shorter than the average inter-nuclear separations; the iron nuclei may be regarded as forming classical ionic plasmas.

Over the bulk of the crustal parts, the nuclei are considered to form a *Coulomb solid*. A neutron star may then be looked upon as consisting of an ultra-dense interior of neutron fluids with fractional constituents of protons and electrons, a crust of Coulomb solids, and a thin layer of "ocean" fluids. Electron transports and photon opacities in the outer crust and in the surface layer play the crucial parts (Gudmundsson, Pethick, & Epstein, 1982) in the estimate of the cooling rates for neutron stars (Nomoto & Tsuruta, 1981).

10.3.2 NON-RADIAL OSCILLATIONS

Non-vanishing shear moduli associated with the crustal solids (Fuchs, 1936; Ogata & Ichimaru, 1990) lead to a prediction of rich spectra in the oscillations of a neutron star. In conjunction with such structures, McDermott et al. (1985, 1988) were the first to analyze non-radial oscillations of neutron stars, with the predictions of the bulk and interfacial modes, associated with the non-vanishing shear moduli of the crustal solid, with characteristic periodicity on the order of milliseconds.

First-principles calculations of shear moduli for Monte-Carlo-simulated Coulomb solids, with inclusion of the Coulomb glasses of Sec. 4.6, were presented (Ogata & Ichimaru, 1989, 1990); the results have been applied for improved analyses of the non-radial oscillations (Strohmayer et al., 1991).

10.3.3 CENTRAL CORE

In the central core of a neutron star with a mass density in excess of 1 billion ton/cm^3, above the discontinuous deconfinement transitions, a phase with the *quark–gluon plasmas* is expected (Baym, 1995; Yagi, Hatsuda, & Miake, 2005).

11

PLASMA PHENOMENA AROUND NEUTRON STARS AND BLACK HOLES

The radio astronomy commenced by Jansky's discovery of cosmic radio waves in 1931 has achieved remarkable progress thanks to the subsequent development in radio technology. In 1967, Hewish, Bell, and their collaborators in England discovered in our Galaxy astronomical objects emitting periodic radio pulses some tens of a millisecond in width at a regular interval of approximately one second; the objects were called *pulsars*. In 1968, a pulsar with a pulse width of 2 ms and a period of 33 ms was discovered at the center of the Crab Nebula, the supernova remnant of the year 1054. Figure 11.1 shows the Crab Nebula observed by the Hubble Telescope (NASA); indicated by the arrow is the location of the pulsar.

The age of artificial satellites initiated by the launching of Sputnik in 1957 opened up our "eyes" to cosmic X-rays, which cannot be detected on the earth because of atmospheric absorption. In 1962, a "bright" *X-ray star* was discovered in the constellation of Scorpius (Sco X-1), emitting X-rays with luminosity greater than a thousand times the total luminosity of the Sun. In 1971, X-ray pulsars emitting pulsed X-rays at regular intervals were discovered in the constellations of Centaurus and Hercules (Cen X-3 and Her X-1). In the same year, an intense X-ray object with a rapid temporal variation was discovered in Cygnus (Cyg X-1), which has subsequently been identified as the first stellar black hole in the Galaxy.

Those conspicuous astrophysical phenomena are produced, in fact, through the radiative processes in plasmas around neutron stars or black holes. A neutron star and a black hole are those states of stars expected in their final stages of evolution (cf., Figure 1.2). In this chapter, we shall elucidate how those stars with all their nuclear fuels exhausted can produce such remarkable radiation.

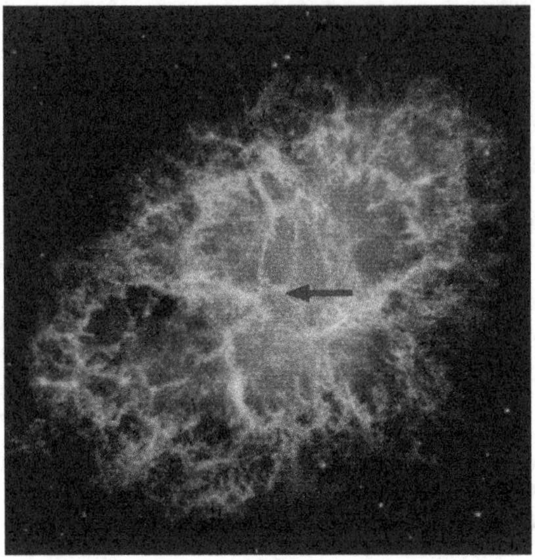

FIGURE 11.1 Crab Nebula observed by Hubble Telescope (NASA). Indicated by the arrow is the location of the Crab Pulsar.

11.1 PULSARS

As remarked in the opening of this chapter, the discovery of pulsars in 1967 was indeed an epoch-making event that brought home sensational developments in plasma physics as well as in nuclear physics.

11.1.1 DISCOVERY

Figure 11.2 shows the record of pulsar discovery reported in Hewish et al. (1968). This is the pulsar now known as PSR 1919+21*; it emits radio pulses (80.5 MHz and 81.5 MHz in Figure 11.2) with a repetition period of 1.3373 s and an average width of 25 ms. Since then until the summer of 1980, some 328 such pulsars were discovered in the Galaxy.

11.1.2 CHARACTERISTIC FEATURES

Some of the typical characters exhibited in those radio pulsars (e.g., Manchester & Taylor, 1977; Michel, 1982) are summarized in the following:

1. The average period (P) and pulse shape of radio pulsars are very stable.
2. The distribution of pulsar periods is peaked somewhere around one second; the largest period known is approximately four seconds.

* The prefix PSR is abbreviation of "pulsar." A four-digit number following PSR indicate right ascension (in 1950.0 coordinates); a sign and two digits are added to designate degrees of declination.

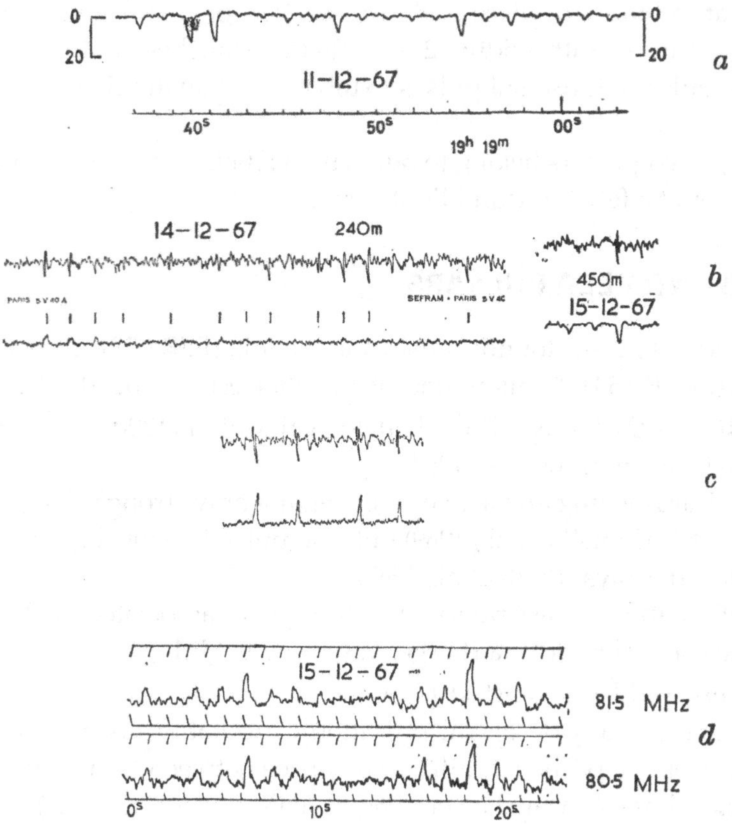

FIGURE 11.2 Record of pulsar discovery on 11–15 December 1967 (Hewish et al., 1968).

3. The repetition period increases gradually over time. Denoting such a rate of increase by $\dot{P}$ (= dP/dt),

$$\tau \equiv \frac{P}{2\dot{P}}. \tag{11.1}$$

takes on 10^6–10^8 years for most of the pulsars. This τ thus gives a rough measure on the "age" of pulsars.

4. The radio wave emitted is intense; an equivalent temperature fitted to the Rayleigh–Jeans law reaches 10^{21}–10^{30} K. This simply means that the radio wave is non-thermal.
5. The average pulse width is approximately proportional to the period; the pulse width ranges 4°–20° when the pulse period is taken to be 360°.
6. With a few exceptions, the pulsar radio waves are weakly polarized (Manchester, Taylor, & Huguenin, 1975). When a polarization is observed, a linear polarization is predominant, and in some cases, circular polarizations are observed.

7. Looking at individual pulses with a resolution of ~1 ms, one can distinguish various subpulses with widths 2°–5° (Taylor, Manchester, & Huguenin, 1975). For some pulsars, those subpulses exhibit a systematic drift across the pulse window.
8. Those observed pulsars belong to our Galaxy; their distances from the earth are on the order of a few thousand light years.

11.1.3 CRAB AND VELA PULSARS

As for the notorious pulsar found in the Crab Nebula (PSR 0531+21), we add the following: Its period $P = 33.098$ ms is one of the shortest among the known pulsars; $\dot{P} = 4.22689 \times 10^{-13}$ is the largest. The calculated value of τ is 1200 years, which roughly corresponds to its birth in the year 1054.

The Crab Pulsar is young and active, emitting not only strong radio pulses (Staelin & Reifenstein, 1968; Comella et al., 1969) but also pulsed visible light (Cocke, Disney, & Taylor, 1969) and X-rays (Bradt et al., 1969).

PSR 0835-45 is another fast pulsar which appears associated with a supernova remnant in the constellation Vela (Large, Vaughan, & Mills., 1968). Its characteristics are: $P = 89.206$ ms, and $\dot{P} = 1.25264 \times 10^{-13}$.

In addition to the steady decay, the pulsation period of the Vela pulsar exhibited a sudden decrease of two parts per million sometime between February 24 and March 3, 1969 (Radhakrishnan & Manchester, 1969; Reichley & Downs, 1969).

11.2 ROTATING MAGNETIC NEUTRON STARS

When the discovery of pulsars created a wake of sensation, one of the first questions asked, naturally, was:

11.2.1 WHAT ARE THE PULSARS?

Candidates for the astronomical objects, which can sustain short-term variations ranging from a few milliseconds to a few seconds, are confined to those compressed objects, such as a white dwarf and a neutron star, expected in the final stages of stellar evolution (cf. Figure 1.2).

As for the characteristic motion, the stellar rotation and vibration, as well as the orbital motion of a binary system, were considered; the last one was soon discarded because of the lifetime considerations due to the emission of gravitational waves.

A white dwarf is formed in the balance between the gravitational self-contraction and the Fermi pressure stemming from degenerate electrons (cf. Sec. 6.1.2). Typically, its mass is about the solar mass ($M_S \cong 1.99 \times 10^{30}$ kg) and the radius is approximately one percent of the solar radius ($R_S \cong 6.69 \times 10^5$ km).

A calculation on the basis of such a model reveals that the period of radial oscillations in its fundamental mode cannot become smaller than 2 s. A consideration on the basis of the escape velocity of matter from the surface of a rotating star indicates that a theoretical lower limit of the rotational period is approximately 1 s. Those numbers are quite inconvenient in explaining the observed features of the pulsars.

It was also noted that none of the observed pulsars were associated with existing white dwarfs.

The neutron star is a final state of a stellar object, which may be created in the supernova remnant; it is even more compressed than the white dwarf. Typically, one assumes a neutron star of a solar mass with a radius of approximately 10 km; the density in its central domain reaches 10^{14}–10^{15} g/cm^3. The period for the radial vibration of such a neutron star is estimated to be 10^{-3}–10^{-4} s, which appears too small to account for the pulsar observation. The lower limit of the rotational period, however, turns out 10^{-2} s, which adequately explains the pulse repetition periods of pulsars, including the Crab.

The observed association of the Crab and Vela pulsars with supernova remnants has been regarded as a positive evidence for the neutron star model of pulsars. In addition to Crab and Vela, the following pulsars are known with their supernova association: PSR 1919+21 (Goss & Schwarz, 1971), PSR 1154-62 (Large & Vaughan, 1972), PSR 0611+22 (Davies, Lyne, & Seiradakis, 1972), PSR 2021+51 (Verschuur, 1973), and PSR 1919+14 (Hulse & Taylor, 1974).

That the signals from PSR 0950+08 contain rapid impulses of 175 μs and are frequently accompanied by variations of ~10 μs (Rickett, Hankins, & Cordes, 1975) is likewise a favorite piece of evidence for the neutron star hypothesis. The total extent of the radio source would at most be the distance over which radio waves propagate in the period of rapid variation. Distances over which the radio waves propagate in 10–175 μs are 3–50 km. These are about the same order of magnitude as a typical neutron star radius of 10 km but are substantially smaller than a radius of the white dwarf.

All the evidence accumulated above appear to point to the neutron stars as the likely model for the pulsars.

11.2.2 PULSAR MAGNETIC FIELD

It is expected that a neutron star is accompanied not only by a rotational motion but also by a strong magnetic field. Consider a situation in which a star similar to the Sun with an average surface magnetic field of 10^2 gauss and a rotational velocity of 10^{-7} rad/s is compressed by a factor of 10^{-5} in radius and thereby becomes a neutron star. Assuming also that the total magnetic flux and the total angular momentum are both conserved in the process of contraction, we estimate that the resulting neutron star would have a surface magnetic field of approximately 10^{12} gauss and a rotational angular velocity of 10^3 rad/s.

This value of rotational velocity, of course, far exceeds the repetition frequency of pulsars now observed. The rotational energy may, however, be lost rather quickly in

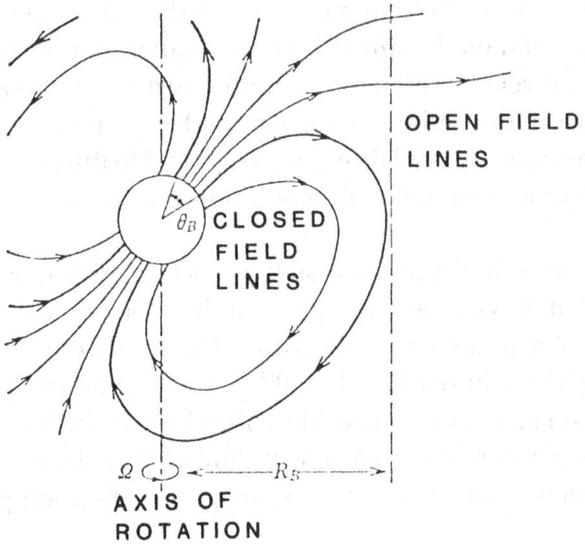

FIGURE 11.3 Rotating magnetic neutron star.

contraction; the neutron star so created may begin rotating with a reduced velocity, accounting for the present observation.

11.2.3 SPINNING DOWN OF PULSARS BY MAGNETIC DIPOLE RADIATION

Under the assumption that the pulsars are the rotating magnetized neutron stars (see Figure 11.3) spinning down by emitting magnetic dipole radiation (Ostriker & Gunn, 1969; Pacini, 1967), we may formulate the rate of change in their angular velocity Ω $(=2\pi/P)$ as given by

$$\frac{d\Omega}{dt} = -\frac{2}{3c^3}\frac{R^6B^2\sin^2\alpha}{I}\Omega^3, \tag{11.2}$$

where I ($\sim 10^{45}$ g·cm^2) is the moment of inertia of the neutron star and α is the angle between the magnetic dipole and rotation axes (Chanmugam, 1992).

If, for the sake of simplicity, we assume that $\alpha = \pi/2$, the magnetic field of the neutron star may then be inferred to be

$$B = \frac{\sqrt{6c^3 I\, P\dot{P}}}{4\pi R^3}. \tag{11.3}$$

Figure 11.4 shows various pulsars plotted in a B versus P diagram. Based on the dipolar radiation hypothesis, we here find that the magnetic fields have been inferred to be 10^{11} gauss $\lesssim B \lesssim 10^{13}$ gauss for most of the pulsars.

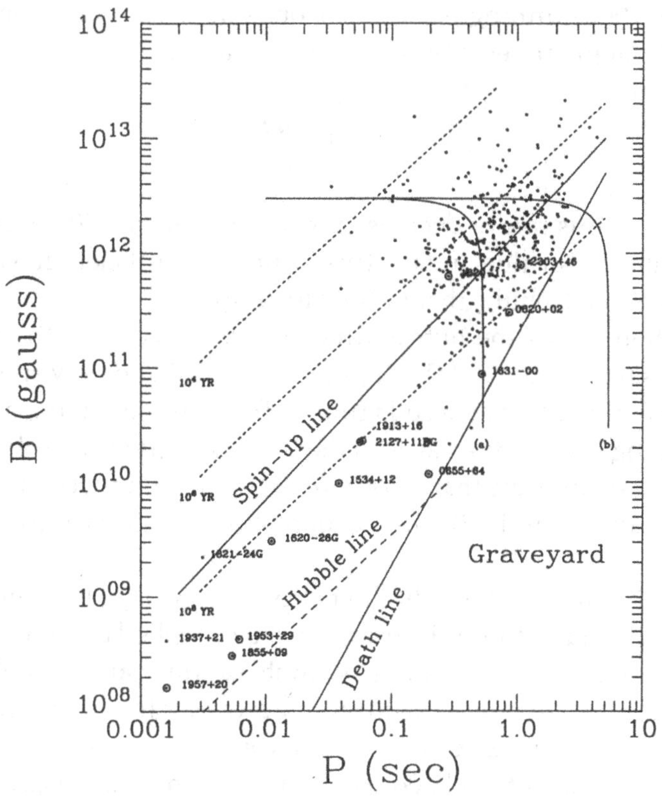

FIGURE 11.4 Magnetic fields of pulsars as a function of their periods, assuming that spinning down results from emission of magnetic radiation. If pulsars are born with short periods, and their magnetic fields are constant, they evolve from left to right and reach the dashed lines at the ages (in years) indicated. When pulsars cross the death line, they die and enter the "graveyard." Pulsars in binaries are indicated by a dot with a circle; those in a globular cluster by "G" (Chanmugam, 1992).

Note, however, that the deduced field strengths would be different if the spin-down was due to higher multipole radiation (Krolik, 1991). More importantly, the pulsar's electromagnetic fields may act to polarize the surrounding medium so that the neutron star is actually situated in a magnetospheric plasma (Goldreich & Julian, 1969); a simple hypothesis of radiation in vacuum may not be applicable.

In the latter connections, we must particularly point to the strong influence that the Crab Pulsar exerts on the Crab Nebula as a whole.

11.2.4 SPINNING DOWN OF CRAB PULSAR AND THE CRAB NEBULA ACTIVITIES

The rotation energy

$$E_R = \frac{1}{2} I\Omega^2 \tag{11.4}$$

associated with the spinning of a neutron star with the moment of inertia I (~10^{45} g·cm^2) and angular velocity Ω decreases at a rate

$$\frac{dE_R}{dt} = I\Omega\frac{d\Omega}{dt}. \tag{11.5}$$

For the Crab Pulsar, this rate of decrease is computed to be ~10^{39} erg/s, based on the observational data. This value, in fact, turns out to be sufficient in supplying all the luminosity of the Crab Nebula, that is, ~4×10^{38} erg/s.

This observation solves the outstanding puzzle concerning the activities of the Crab Nebula. It is the remnant of the supernova in 1054 and now spreads over a spatial extent of several light years (see Figure 11.1). From the region, a wide spectrum of radiation ranging from radio waves to the γ-ray is emitted. The observed radiation characteristics clearly indicate that those emissions are non-thermal, produced by the *synchrotron radiation*, that is, by the interaction of relativistic electrons with magnetic fields.

The estimated lifetimes of those high-energy electrons are far shorter than the age of the Crab Nebula, and so those electrons must be supplied continuously by external sources. A pulsar having been discovered in its central part, it would be natural to assume that the rotating magnetic neutron star provides the creation and acceleration mechanisms necessary for the relativistic electrons.

It may thus give us a sort of elated feeling to imagine that a heavy "toy top" with a radius of ~10 km spinning at a high speed of 30 revolutions per second releases its rotational energy gradually and thereby "energizes" the entire nebula spreading over an extent of several light years. The X-ray image of the Crab Pulsar in Figure 11.5 may graphically substantiate such features.

We may also assume that the strong magnetic field of a neutron star plays the central role in converting the rotational energy to the radiation. For example, the rotational energy may in part be converted to magnetic dipole radiation, as we remarked in conjunction with Figure 11.4. The frequency of such a magnetic radiation, however, is simply P^{-1}; such radiation by itself would not produce the pulses.

11.2.5 CONSTRUCTING THE RADIO BEAMS

Let us then proceed to consider the pulsar radiation mechanisms by assuming a fundamental model of a rotating neutron star with a surface magnetic field on the order of 10^{12} gauss, as in Figure 11.3 (e.g., Manchester & Taylor, 1977; Michel, 1982).

At this stage, we recall the stability of the average pulse shape as remarked in Sec. 11.1.2 (1). Combining the intensity of the radio wave observed and the estimate of the pulsar distances, we can infer the energy density of the radiation field in the vicinity of a pulsar. For stability in mechanical terms, this energy density must be substantially smaller than the energy density of the plasmas that produce the radiation. The only conceivable source to provide such mechanical stability is the strong magnetic

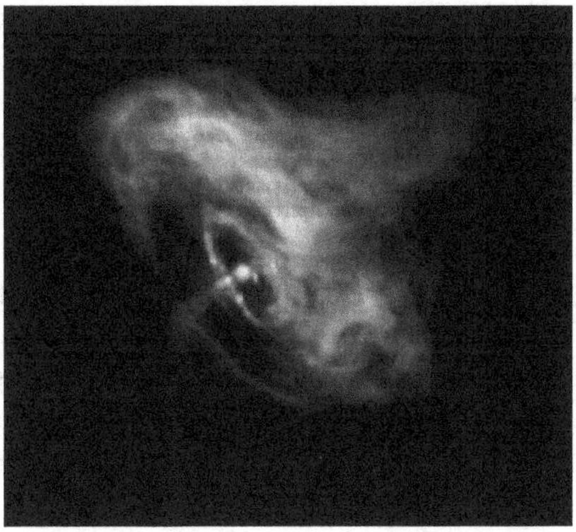

FIGURE 11.5 X-ray image of the Crab Pulsar by Chandra X-ray Observatory (NASA).

field associated with the neutron star. Accordingly, we may accept a model in which the radio waves are emitted near the stellar surfaces.

A simple mechanism to establish radio pulses with a rotating star is to assume a searchlight beam of radio wave around the axis of rotation. Under these circumstances, we may observe the radio wave only when the beam is directed toward us; thus, the opening angle of the searchlight beam is geometrically connected to the effective widths of the pulses.

As we observe in Figure 11.3, some of the magnetic lines of force streaming out from the surface of the rotation star may return to the stellar surface, while some others may stream out into the interstellar space. We note, on the one hand, those charged particles with the closed magnetic lines of force cannot be accelerated effectively by an external force; on the other hand, the acceleration is rather efficient in the space with open field lines. Consequently, the boundary between the open and closed field lines may determine the opening angle of the searchlight, where radio emission is predominant.

Let R_B be the largest radius where the closed magnetic lines of force can reach (cf. Figure 11.3). Assuming a dipolar magnetic field, we may estimate the opening angle of the searchlight as

$$\theta_B = 3\left(\frac{R}{R_B}\right)^{1/2} \tag{11.6}$$

where R is the radius of the neutron star.

Two theoretical possibilities have been suggested for R_B. One is to take

$$R_L = \frac{c}{\Omega}, \tag{11.7}$$

where the rotational velocity becomes equal to the light velocity. The other is to take R_c, where the gravitational force balances with the centrifugal force (Roberts & Sturrock, 1973), that is,

$$\frac{GM}{R_c^2} = \Omega^2 R_c. \tag{11.8}$$

Here $G = 6.6720 \times 10^{-8}$ erg·cm·g^{-2} is Newton's gravitational constant and M is mass of the neutron star.

This radius, being an order of magnitude smaller than R_L, gives a slightly larger estimate of θ_B. The value of θ_B determining the ratio between pulse widths and periods in these models appears to give a fairly consistent estimate vis-à-vis the observed data.

11.2.6 CREATING THE PLASMAS

The remaining issues in conjunction with a model setting are first to apply an accelerating field in the space specified previously, then to create plasmas emitting the radiation, and finally, to compare the results with observation. These are complex issues. At this stage, we do not know how to create such plasmas, not to mention their properties.

Sturrock (1971) and later Ruderman and Sutherland (1975) proposed a theoretical model accounting for a radiation mechanism of pulsars with the inclusion of the issues of plasma production. To a degree, it physically resembles the processes in ordinary gaseous discharges. One evokes the electron–positron pair production arising from the interaction between the γ-ray and the magnetic field, in place of the electric discharges of atoms and molecules in gases. The threshold energy $2mc^2$ for the pair production thus plays the part of the ionization potential.

To begin, we note that a neutron star may be viewed as a perfect conductor, because it contains a substantial amount of conduction electrons and protons (cf. Sec. 1.1.1 and Figure 1.3). Generally, in the rotating magnetic neutron star, the stellar matter would move relative to the magnetic field, which in turn would exert a force onto the stellar matter. In a perfect conductor, the net force acting on a charged particle should vanish; a space-charge distribution is thus created inside the star just to cancel the effect due to the magnetic force. In other words, a rotating magnetic star polarizes itself electromagnetically and thereby produces an internal electric field.

Electric polarization of a star then additionally induces a potential difference in the space exterior to the star. The maximum amount of the potential difference created between the star and the interstellar space along the magnetic lines of force reaches $\Omega\Phi/2\pi c$, where Φ ($\simeq \pi\theta_B^2 R^2 B/4$) is the total number of the open magnetic lines of force in Figure 11.3. The bulk of such a potential difference is moreover concentrated near the stellar surface up to the altitude on the order of the stellar radius. This potential difference would naturally increase with the rotational velocity Ω of the star; for an ordinary pulsar with $P \simeq 1$ s, it takes on a value of around 10^{12} V. This is a potential capable of accelerating an electron up to a super-relativistic energy of γmc^2 with $\gamma \approx 2 \times 10^7$.

Suppose that an electron (or a positron) is placed above the neutron star near a magnetic pole. Since the electric field is extremely strong, such an electron would be accelerated immediately to a super-relativistic ($\gamma \gg 1$ for a particle with energy γmc^2) energy. The relativistic electrons traveling along magnetic lines of force with a finite radius of curvature emit γ-ray photons in its direction of motion with a characteristic energy determined by the radius of curvature and the energy value; it is the synchrotron radiation.

When the energy of such a photon exceeds $2mc^2$, a possibility of creating an electron–positron pair arises through interaction with adjacent magnetic lines of force. The mean free path of a photon for the pair production depends on the strength of the magnetic field, the curvature, and the photon energy; it tends to decrease as each of those quantities increases.

If the effective length of acceleration is comparable to the mean free path, then a pair production will take place. Electron and positron so produced are then accelerated separately by the electric field and thereby emit photons. Those photons then produce electron–positron pairs, which are then accelerated and emit photons. An avalanche of pair production thus takes place, and electron–positron plasmas may be created, analogous to "spark" discharges in gaseous substances.

11.2.7 A PULSAR EMISSION MECHANISM

If a spark discharge of this nature is maintained near the magnetic pole, electron–positron plasmas may be produced. Since an electron and a positron carry the electric charges of opposite sign, they produce a counter-streaming flow of charged particles in the electric field. The *two-stream instability* of plasma oscillation (e.g., Buneman, 1959) may then take place; spatial bunching of charged particles may thus develop (Ichimaru, 1970). In Sec. 3.5, we treated excitations of the ion-acoustic wave due to drift motion of the electrons relative to the ions; the resultant enhancement of density fluctuations, reminiscent of the critical opalescence in the vicinity of a liquid-gas phase transition, was considered. The bunching of charged particles under present circumstances may resemble these phenomena.

The bunches of charged particles so produced travel along magnetic lines of force and thereby emit electromagnetic waves by the curvature acceleration. If the sizes of bunches are smaller than those wavelengths in the radio domain, the N charged particles in a bunch may emit the waves with coherent phases. Since the strength of electromagnetic radiation is proportional to the square of the electric charge, it may become N times greater in a coherent situation than in an incoherent one. Since N can take on an extremely large number, the high effective temperatures noted in the item Sec. 11.1.2 (4) may be accounted for in terms of such bunching.

As the rotation is slowed down and the effective strength of acceleration decreases, electrons and positrons cannot be accelerated effectively; the high-energy photons for the pair productions cannot be created. In these circumstances, a neutron star even with a substantial magnetic field cannot be observed as a pulsar, because radiating plasmas cannot be maintained. This may be a reason for not finding a pulsar with a long period, as noted in the item Sec. 11.1.2 (2); Figure 11.4, in fact, shows the scarcity of pulsars with $P > 4$ s.

The foregoing is a scenario explaining the mechanism of a pulsar. Some other possibilities and alternatives have also been considered and examined (e.g., Manchester & Taylor, 1977; Michel, 1982).

11.3 X-RAY PULSARS

Except for a few examples (e.g., PSR 0531+21), all the pulsars treated in the previous sections are radio pulsars. In 1971, X-ray pulsars were discovered in the constellations of Centaurus and Hercules, and they were named Cen X-3 and Her X-1 (Giacconi et al., 1971; Tananbaum et al., 1972a). Figure 11.6 shows the X-ray signal from Her X-1. The pulse repetition period was 4.84 s for Cen X-3 and 1.24 s for Her X-1.

11.3.1 CLOSE BINARY SYSTEMS

It was also confirmed that those X-ray pulsars are members of binary stars. Members of the binaries perform Keplerian orbital motion with respect to their centers of

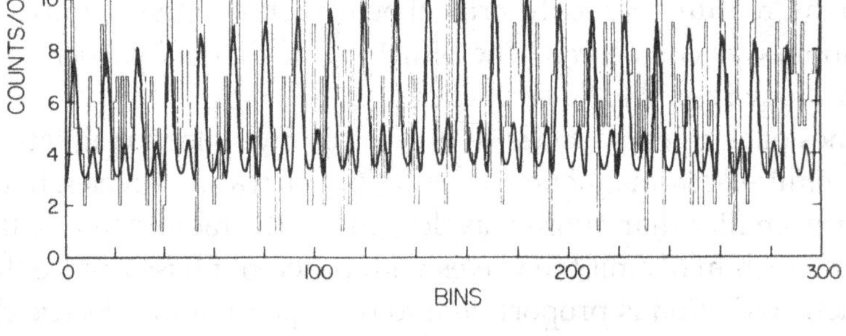

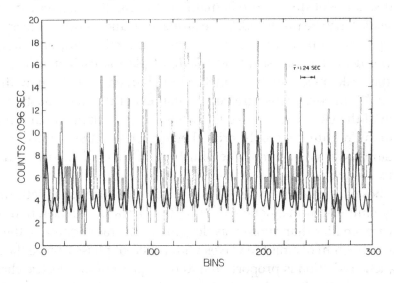

FIGURE 11.6 The counts accumulated in 0.096-second bins from Hercules X-1 during the central 30 seconds of a 100-second pass on 6 November 1971. The heavier curve is a minimum χ^2 fit to the pulsations of a sine function, its first and second harmonics plus a constant, modulated by the triangular response of the collimator (Tananbaum et al., 1972a).

gravity. Periods of such an orbital motion observed were 2.1 days with Cen X-3 and 1.7 days with Her X-1. Note that those are far shorter than the orbital period of the Earth (approximately 365 days), for example.

Orbital motion was confirmed by the fact that the pulse arrival times were not evenly distributed (see Figure 11.7). This is the Doppler effect treated in Sec. 3.1.2. While an X-ray pulsar in a binary was approaching us, interpulse separations shorter than the average were observed; while receding, the separations became wider. The orbital data can thus be decoded from the arrival times. The deviation shown in Figure 11.7 follows a sinusoidal curve, implying the pulsar performing almost a circular orbital motion with negligible eccentricity.

Another remarkable feature in Figure 11.7 is the fact that the X-ray pulses are unobservable at some phases of the orbital motion. This means an "eclipse" when a small X-ray star travels behind a giant star with a radius several times that of the Sun. Such an eclipse thus provides another piece of evidence that the X-ray pulsar is a member of a binary system.

Analyzing the pulse arrival times in terms of the Kepler's third law reveals that the distances between the binaries take on small values near 10^7 km, which are only an order of magnitude larger than the solar radius. Such is therefore called a *close binary system* (CBS).

11.3.2 ACCRETION

Let us now take a system of coordinates fixed onto such a binary system and consider a force acting on a test mass placed on this system. First, it is attracted to each star by the gravitational force; second, it is repelled from the center of mass for the binary by the centrifugal force. When the test mass is located far from the binary, the centrifugal force wins; it is ejected. When the test mass is placed near the binary, the gravitational attraction wins.

We may thus define effective potential and equipotential surfaces in the combined field of the centrifugal force and the gravity. The *Roche limit* may thus be defined in

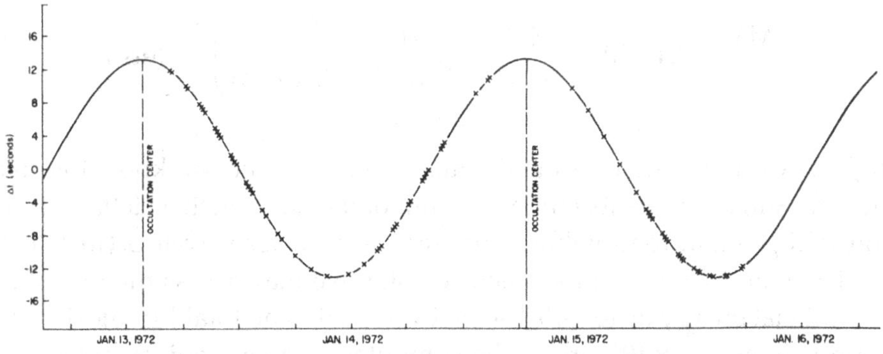

FIGURE 11.7 The difference Δt between the time of occurrence of a pulse and the time predicted for a constant period is plotted as a function of time (Tananbaum et al., 1972a).

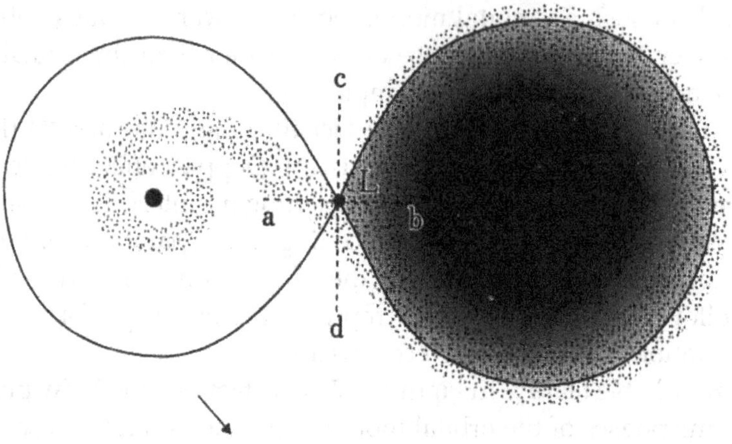

FIGURE 11.8 Roche limit and accretion in the close binary system.

Figure 11.8 as the smallest of those equipotential surfaces that enclose both stars. The effective potential takes on a maximal value at L along the line ab and a minimal value at L along the line cd; the point L thus makes a saddle point.

If the right-hand star in Figure 11.8 has a size comparable to or greater than the Roche limit, then the atmospheric matter, or plasmas, may flow over the Roche limit through the saddle point L, falling into the gravitational domain of the left-hand star. If the left-hand star is a dense degenerate star such as a neutron star (or a black hole), a sizable amount of gravitational energy is liberated as the overflowing matter falls onto the stellar surface; a copious emission of X-ray is thus expected.

The process described above is called *accretion*. Let us calculate the amount of gravitational energy liberated through such an accretion. We denote the radius of the accreting star by R, the mass by M, the rate of mass accretion per unit time by $\dot{M}$. Since the accreting matter eventually sticks onto the star, it means an increase in the stellar mass. The rate of energy liberated then is

$$\frac{GM\dot{M}}{R} = 8.4\times 10^{36}\left(\frac{M}{M_S}\right)\left(\frac{\dot{M}}{10^{-9}M_S\,/\,\mathrm{yr}}\right)\left(\frac{R}{10\,\mathrm{km}}\right)^{-1} \quad (\mathrm{erg/s}). \tag{11.9}$$

Although we cannot predetermine the mass accretion rate, we know the solar wind by which the Sun ejects a substantial amount of its mass. Such a stellar wind being a common affair, we may reasonably assume a mass transfer at a rate of $(10^{-9}$–$10^{-8})\,M_S$/yr. Taking $M = M_S$ and $R = 10$ km for a neutron star, we may assess the rate of releasing the gravitational energy at 10^{37}–10^{38} erg/s from (11.9). The total luminosity of the Sun being approximately 4×10^{33} erg/s, the computation above points to the possibility of devising an X-ray star "brighter than a thousand Suns" through accretion onto a neutron star in a CBS.

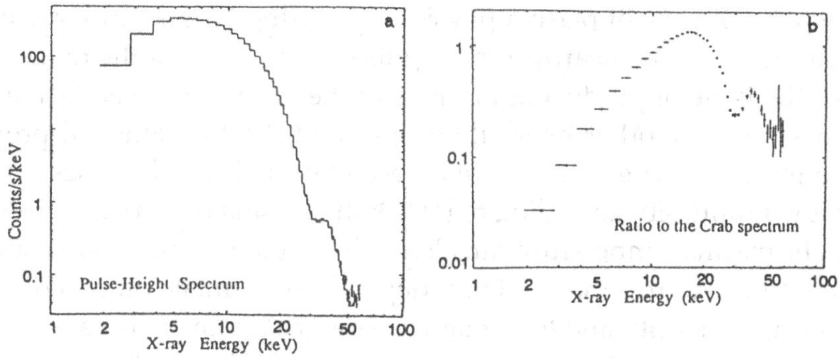

FIGURE 11.9 The prominent CRSF observed at 28.5 keV from the transient X-ray pulsar X0331+53 with *Ginga* (Makishima et al., 1990). (a) The observed raw pulse-height spectrum, fitted with an empirical model for cyclotron resonance. (b) The spectrum normalized to that of the Crab Nebula.

11.3.3 CYCLOTRON RESONANCE SCATTERING FEATURE

The neutron star onto which plasmas accrete may carry a strong magnetic field, as with the case of the radio pulsars. In the present case, detection of the cyclotron resonance scattering feature (CRSF), that is, a characteristic feature in the observed X-ray spectrum due to electron-cyclotron resonance scattering offers a reliable technique of estimating the field strength in the accretion column above the magnetic pole. In terms of the observed cyclotron resonance energy E_c, the field strength is given by

$$B(10^{11}\text{gauss}) = 0.862\, E_c\ (\text{keV}). \tag{11.10}$$

Figure 11.9 exhibits the best example of CRSF as observed by the *Ginga* satellite, a Japanese X-ray astronomy satellite launched in 1987 (Makishima, 1995). The feature observed at 28.5 keV implies the field strength of 2.5×10^{12} gauss for the transient X-ray pulsar X0331+53. We might additionally remark that this technique has been applied to Her X-1, in which a CRSF was seen either at ~35 keV in absorption, or at ~60 keV in emission (Trümper et al., 1978; Voges et al., 1982).

11.3.4 ACCRETION MODEL OF X-RAY PULSARS

The accretion of plasmas in a CBS is one of the fundamental models accounting for X-ray pulsars such as Cen X-3 and Her X-1 (Pringle & Rees, 1972; Davidson & Ostriker, 1973; Lamb, Pethick, & Pines, 1973; Shapiro & Teukolsky, 1983). Plasmas falling toward the neutron star may form an accretion disk, to be accounted for in Sec. 11.4, around a magnetized neutron star. Plasmas forming an accretion column falls onto the magnetic poles. Through accretion, the plasmas are strongly heated in the gravitational field and thereby emit X-ray from the pole spots. Emitted X-ray is modulated by the stellar rotation; periodic pulse signals, as shown in Figure 11.6, are thus observed.

From the point of view of plasma physics, the setting around an X-ray pulsar differs significantly from that nearby a radio pulsar. In the case of the radio pulsar, we had to solve the issue of producing plasmas at the onset. In the accretion model of an X-ray pulsar, on the other hand, the origin and the fundamental properties of plasmas are more or less given. The mass accretion rate can be assessed from the observed X-ray luminosity according to (11.10); the plasma density may be estimated therefrom. The plasma temperature can also be inferred from the energy spectrum of the observed X-ray. We thus have a fairly dependable estimate on two of the fundamental quantities—density and temperature—describing the plasma.

On the basis of those estimates, one may proceed with analyzing the hydrodynamic behavior of the accreting plasma, interaction with the stellar magnetic field as well as heating and X-ray emission in the vicinity of the magnetic poles.

11.4 BLACK HOLE MODEL OF CYGNUS X-1

In 1916, the year after Albert Einstein laid down the final formulation of the field equations of general relativity, Schwarzschild published a solution for the field equations (Schwarzschild, 1916) that was later understood to describe a black hole (Finkelstein, 1958; Kruskal, 1960). In 1971, another X-ray star, Cyg X-1, attracted the attention of many investigators (Oda et al., 1971; Rappaport, Doxsey, & Zauman, 1971; Schreier et al., 1971), which has since been accepted as a first stellar black hole observed in the Galaxy.

Figure 11.10 particularly exhibits the X-ray signal from Cyg X-1 with violent short-term variability. It does not, however, contain well-defined periodic components as in the case of Cen X-3 and Her X-1; the latter did contain periodic components as well as rapid variability. The X-ray intensity of Cyg X-1 varied with a time constant of less than a second. Detailed measurements further revealed X-ray bursts of a few hundred microseconds.

It has also been confirmed that Cyg X-1 is a member of a CBS (cf. Sec. 11.3.1) with an orbital period of 5.6 days. The companion of Cyg X-1 is an optically identified, ordinary star, HDE226868, with absorption lines showing Doppler effects due to orbital motion. The emission lines observed from the matter accreting onto Cyg X-1 likewise exhibited the Doppler effects, which varied in opposite phases.

11.4.1 ENERGY SPECTRA AND VARIABILITY OF X-RAY EMISSION

Figure 11.11 shows salient features observed in the energy spectrum of X-rays (Tananbaum et al., 1972b). The abscissa and ordinate represent the X-ray energy and the count numbers of X-ray photons (per unit area and per unit time), both plotted in logarithmic scales. Until early April of 1971, the Cyg X-1 spectrum was that shown by the kink line in Figure 11.11. Since the energy emission rate is the product of photon energy and count number, the data imply a two-peak structure in the vicinity of a

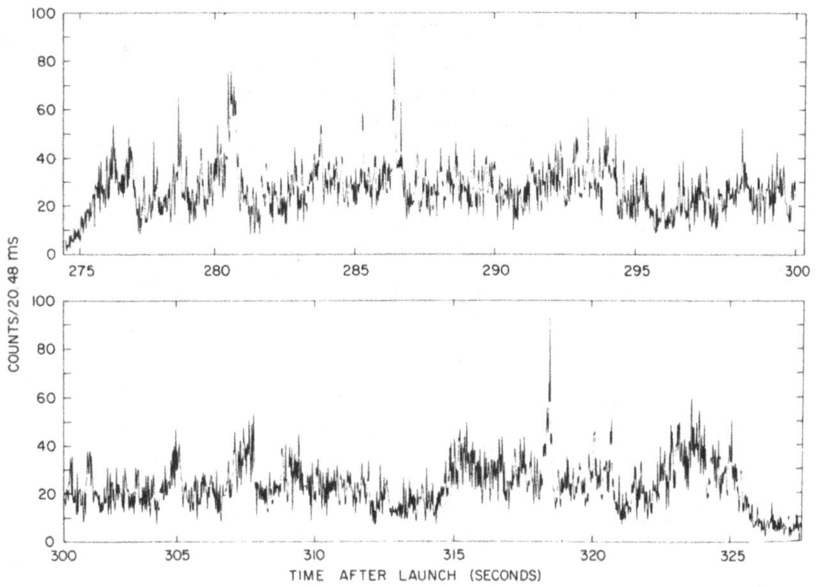

FIGURE 11.10 The entire exposure to Cygnus X-1 as a function of time after launch on 4 October 1973. The count rates are binned every 20.48 ms (Rothschild et al., 1974).

few keV as well as near a few tens of a keV in its spectral distribution. For brevity, we shall call this highly luminous double peak state the "high-mode."

Around early April of 1971, however, the two-peak structure changed into a single-peaked structure represented by the straight line in Figure 11.11 in a period of approximately one day; the peak in the vicinity of a few keV then disappeared (Tananbaum et al. 1972b). This mode of spectrum remained until April 1975, when it returned gradually to the two-peak structure over several days in late April (Holt et al., 1975). In early May, Cyg X-1 again assumed a single peak mode in a transition time of approximately one day (Sanford et al., 1975). Since then, several transitions between those two modes have been reported. For brevity, we shall call this lowly luminous single peak state the "low-mode."

To summarize the observed features of Cyg X-1, we note an intense X-ray emission ($\gtrsim 10^{37}$ erg/s) with rapid time variability (Figure 11.10), the existence of two "modes" in the energy spectrum (Figure 11.11), and transitions between those two modes.

11.4.2 MASS ESTIMATE

It was postulated that Cyg X-1 was a black hole in a CBS. For the energy source of the X-ray emission, an accretion process analogous to that described in Figure 11.8 has been considered.

A black hole is a state of a star compressed further beyond the state of a neutron star within a characteristic radius, called the *Schwarzschild radius,*

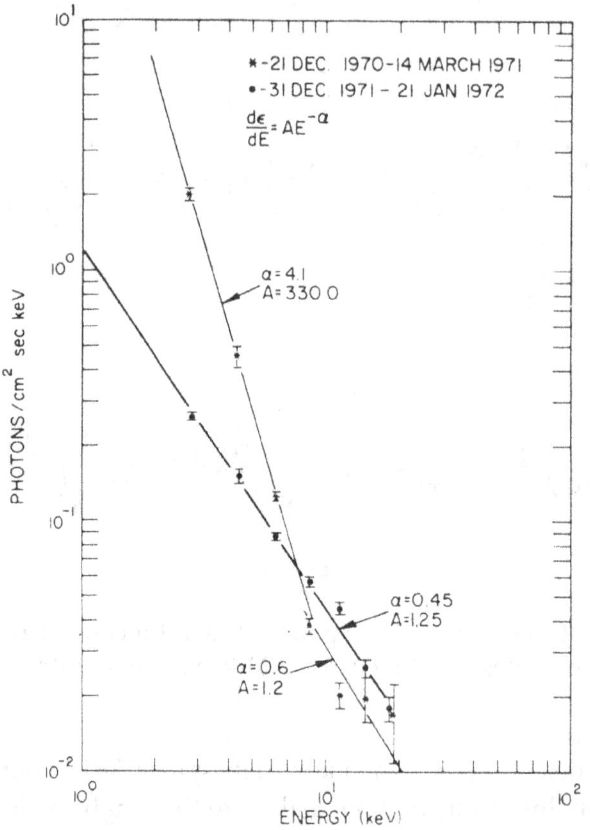

FIGURE 11.11 The average spectra for Cygnus X-1 from before and after the March and April 1971 transition. The spectra are plotted in photons $(cm^2 \cdot s \cdot keV)^{-1}$ versus energy; various power law fits are labeled with the energy index and normalization coefficient (Tananbaum et al., 1972b).

$$R_B \equiv \frac{2GM}{c^2}, \tag{11.11}$$

where G is Newton's gravitational constant and M is the stellar mass (e.g., Weinberg, 1972; Shapiro & Teukolsky, 1983). The associated gravitational field is so strong that even a photon is trapped and cannot escape from within this radius. If we take M to be the solar mass M_S, then R_B takes on 3 km, about one-third of a neutron star radius.

One of the principal reasons for the black hole postulation derives from the mass estimate for Cyg X-1. First of all, we know the parameters for the orbital from combination of observed Doppler shifts in the absorption lines and Kepler's laws. From the spectral type of HDE226868, one derives a relation between its radius and mass. Since accreting plasmas have been observed, the companion star may be viewed as filling its Roche limit. The combination of these considerations enables one to determine a possible range of the mass M of Cyg X-1 as $9M_S \leq M \leq 18M_S$ (Bolton, 1975).

Since Cyg X-1 emits X-rays with a rapid short-term variability (see Figure 11.10), it may naturally be assumed to be a compact star. For stability as a stellar body, the white dwarf or the neutron star has its own upper limit in the mass for existence.

For the white dwarf, it is the *Chandrasekhar mass limit,* which takes on $1.44M_S$ (Chandrasekhar, 1935, 1984), as it consists of the ordinary matter described in the phase diagrams such as Figure 6.1. As for the upper limit for the mass of a neutron star, which consists of the nuclear matter in the phase diagrams such as Figure 10.1, it appears that the upper limit falls around $3M_S$. At any rate, for a compact star with a mass greater than $9M_S$ neither a white dwarf nor a neutron star can exist stably; the only conceivable object remaining is a black hole.

11.4.3 PLASMA ACCRETION TO A BLACK HOLE

Let us then set up a scenario as to how the accreting plasmas behave and emit X-ray in a close binary system containing a black hole (Thorne & Price, 1975; Eardley, Lightman, & Shapiro, 1975; Shapiro, Lightman, & Eardley, 1976). The numbers cited below are the model parameters adopted for Cyg X-1 (Ichimaru, 1977).

Let r be the distance from the accreting black hole. General relativistic effects become important for $r \lesssim 10^7$ cm. The Roche limit is located around $r \simeq 10^{12}$ cm. Stellar matter (plasma) overflows the saddle point L in Figure 11.8 and falls into the gravitational domain of the black hole. The accreting plasmas carry angular momenta due to the orbital motion of the binary system; they tend to form a disk in the plane perpendicular to the axis of rotation as in Figure 11.12. This is the *accretion disk* (Pringle & Rees, 1972; Shakura & Sunyaev, 1973; Novikov & Thorne, 1973).

The thickness $2h$ of an accretion disk can be estimated from the condition for equilibrium between the expanding force $\sim P/h$ due to the pressure P and the contracting force $\sim GM\rho_m h/r^3$ due to the gravitational attraction of the black hole; the balance gives

$$h \simeq \frac{r}{v_K}\left(\frac{P}{\rho_m}\right)^{1/2}. \tag{11.12}$$

Here ρ_m refers to the mass density of the plasma and

$$v_K = \left(\frac{GM}{r}\right)^{1/2} \tag{11.13}$$

is the Keplerian rotational velocity of the disk. Such an accretion disk may be formed at distances, $r \lesssim 3\times 10^{11}$ cm.

The angular momentum per unit mass, rv_K, is proportional to $r^{1/2}$. For the plasma to be able to fall toward the black hole, an excess amount of the angular momentum should be released outward. Viscosity in the plasma participates in the process of such transfer of angular momentum; viscosity also implies internal friction and thus acts to heat the plasma (Eardley & Lightman, 1975; Ichimaru, 1976).

The Keplerian velocity, at which the centrifugal force balances with the gravitational attraction, represents a huge amount of kinetic energy for plasmas accreting in the vicinity of a neutron star or a black hole. If we take a proton at $r \simeq 10$ km, a neutron star radius, and $M = M_S$, then we find

$$\frac{1}{2} m_p v_K^2 \simeq 1.1\times10^{-4}\ \text{erg} \simeq 69\ \text{MeV}. \tag{11.14}$$

So, copious X-ray emission as estimated in (11.9) is expected from plasmas accreting onto the surface of a neutron star.

11.4.4 A BLACK HOLE MODEL OF CYG X-1 OBSERVATION

Analyses of angular momentum transfer, viscous heating, and resulting elementary processes in accreting plasmas may put forward the possibility that two distinct modes exist in the accretion disk, as illustrated in Figure 11.12 (Ichimaru, 1977). The distinction between those two modes depends on the rate of accretion at $r \simeq 3\times10^{11}$ cm where the disk is formed.

At this stage, it is useful to analyze the states of disk plasmas in terms of opacities and emissivity of photon. When the plasma is dense so that the mean free path of a photon is much shorter than a linear dimension of the plasma, the system is said to be optically thick; the plasma radiates quite efficiently, almost like a black body. When the plasma is dilute so that the mean free path of a photon is much longer than the system size, it is said to be optically thin; the plasma radiates inefficiently by bremsstrahlung.

When relatively high-density plasmas ($\dot{M} \gtrsim 3\times10^{-8}\ M_S$/yr) are supplied at $r \simeq 3\times10^{11}$ cm where a disk may be formed, a low-temperature disk starts to form as

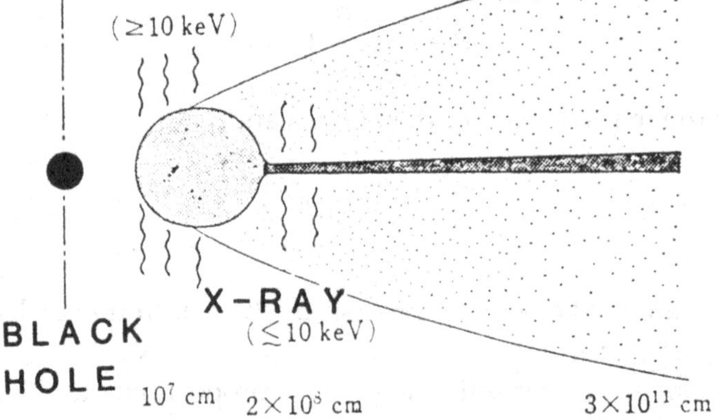

FIGURE 11.12 Schematic cross-sectional view of accretion disks around a black hole, showing a possible bimodal behavior.

photons may be emitted quite efficiently owing to a high opacity of the plasma; heat generated by viscosity is efficiently carried away by radiation. As P/ρ_m is proportional to the temperature, h/r determined from (11.12) takes on a small value; a geometrically thin disk is thus formed.

As the plasma disk gradually falls toward the black hole, an increased amount of the gravitational energy is liberated; the temperature rises. The low-energy peak (~ a few keV) in the kinked spectrum of Figure 11.11 stems from thermal radiation in this vicinity ($r \gtrsim 2\times10^8$ cm).

As the plasma falls further toward the black hole, such a geometrically thin disk becomes thermally unstable in the vicinity of $r \simeq 2\times10^8$ cm. For, if the disk is heated and enlarged, the efficiency of thermal radiation decreases; the thermal energy due to viscous heating cannot be effectively dissipated; the plasma temperature goes further up; and a thermal instability develops therefrom. As a consequence, the disk plasma is abruptly transformed into a high-temperature, geometrically expanded state; its electron temperature may reach $\sim 10^9$ K.

In this domain, photons scattered by relativistic electrons receive substantial enhancement in energies (i.e., frequencies) due to the *inverse Compton processes*. High-energy X-rays with a peak around a few tens of a keV are emitted. The spectrum of radiation represented by the kink line in Figure 11.11 may be accounted for in terms of the evolution of accretion disk as described previously. A double peak structure thus appears, corresponding to the "high-mode."

When relatively low-density plasma ($\dot{M} \lesssim 3\times10^{-8}\ M_S$/yr) is supplied in the vicinity of $r \simeq 3\times10^{11}$ cm, on the contrary, the accretion disk follows an evolution path qualitatively different from that described previously. The rate of bremsstrahlung depends on the frequency of inter-particle collisions, which in turn is proportional to the square of density. Low-density plasmas with scarce inter-particle collisions are inefficient for radiative processes; viscous dissipation goes directly into heating of the plasma. The accreting plasma begins with a high-temperature expanded state.

Although the plasma temperature is high, the X-ray emission is weak in the beginning because of lowness in accretion rate and plasma density. Such an accretion disk, however, joins into the high-temperature expanded plasma in the previous evolutional scenario in the vicinity (10^7 cm $\lesssim r \lesssim 2\times10^8$ cm) of the black hole; it thus produces an X-ray spectrum with a single peak around a few tens of a keV. We thus interpret the "low-mode" represented by the straight line in Figure 11.11 as originating from such an accretion process.

The high-temperature plasma in the vicinity of the black hole ($r \lesssim 10^7$ cm) moves so fast that the plasma may well be in a magnetohydrodynamically turbulent state. Consequently, the observed millisecond bursts may be accounted for in terms of violent motion associated with turbulence; the rapid X-ray variability characterizing Cyg X-1 may thus reflect the special features in the accreting plasmas onto the black hole.

As explained previously, a black hole model has given a consistent explanation for the bimodal characteristics of the Cyg X-1 spectrum. Combining these facts with the mass estimate of Cyg X-1, we may interpret it as highly probable that Cyg X-1 is, in fact, a black hole.

11.5 STELLAR-MASS BLACK HOLES AND SUPERMASSIVE BLACK HOLES

Cyg X-1, treated in the preceding section, is an accreting black hole binary with an estimated mass ranging $9M_S \leq M \leq 18M_S$. Such a system with a stellar-mass black hole is now called a *microquasar.*

In the Universe, there have also been found supermassive black holes with mass ranging $10^6M_S \leq M \leq 10^9M_S$ near the centers of galaxies as *active galactic nuclei* (e.g., Rees, 1988).

11.5.1 MICROQUASARS

Black hole binaries, as we observed with Cyg X-1, have fascinated astronomers, because they go through a cycle of many different activity states.

For example, they can be in a state of high accretion and high luminosity, in which they strongly emit both "soft" (low energy) and "hard" (high energy) X-rays, namely, the bright/soft state; this state explicitly corresponds to the "high-mode" with Cyg X-1. Another state is one of low accretion and low luminosity, in which the hard X-ray emission exceeds that of soft X-rays, namely, the faint/hard state; this state corresponds to the "low-mode" with Cyg X-1.

Figure 11.13 shows schematic illustrations depicting the bright/soft state and the faint/hard state of accretion. As far as physical processes in the accretion disks are concerned, these respectively correspond to the "high-mode" and "low-mode" described in Figures 11.11 and 11.12.

GRS 1915+105 is a $14M_S$ black hole accreting matter from a $0.8M_S$ K3IV star in a wide 33.5-day orbit (Greiner, Cuby, & McCaughrean, 2001). As the first known source

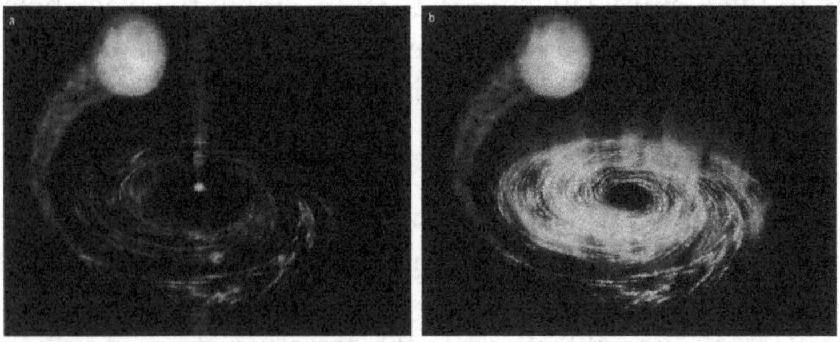

FIGURE 11.13 Illustrative drawings of accretion disks around a black hole, showing possible (a) faint/hard state and (b) bright/soft state. In (a), a narrow relativistic jet of plasma is depicted for radio emission (Proga, 2009).

of superluminal jets in the Galaxy (Mirabel & Rodriguez, 1994), with a light curve exhibiting at least 14 distinct classes of high-amplitude variability due to rapid disk-jet interactions (Eikenberry et al., 1998; Mirabel et al., 1998; Klein-Wolt et al., 2002; Hannikainen et al., 2005), this microquasar provides a fascinating example of coupling between jets and accretion disks around black holes.

To study this relationship, Neilsen and Lee (2009) analyzed the archival High Energy Transmission Grating Spectrometer (HETGS) (Canizares et al., 2005) observation of GRS 1915+105 from the Chandra X-ray Observatory. Between 24 April 2000 and 14 August 2007, the HETGS observed this microquasar 11 times with high spectral resolution, probing 5 of 14 variability classes of this enigmatic X-ray source. Figure 11.14 shows the data, including six observations of the faint, hard, jet-producing state (Dhawan, Mirabel, & Rodriguez, 2000) (observations H1–H6) and five observations of various bright, softer states (observations S1–S5).

11.5.2 SUPERMASSIVE BLACK HOLE IN THE GALAXY

Around the center of our Milky Way Galaxy, a supermassive black hole called Sagittarius A* (Sgr A*) with a mass $\sim 10^6 M_S$ exists, which appears to be accompanied by swirling-around gas clouds.

In Sec. 11.2.4, we took up the spinning down of the Crab Pulsar and its relation with the Crab Nebula activities. The Crab Pulsar is a strongly magnetized rotating

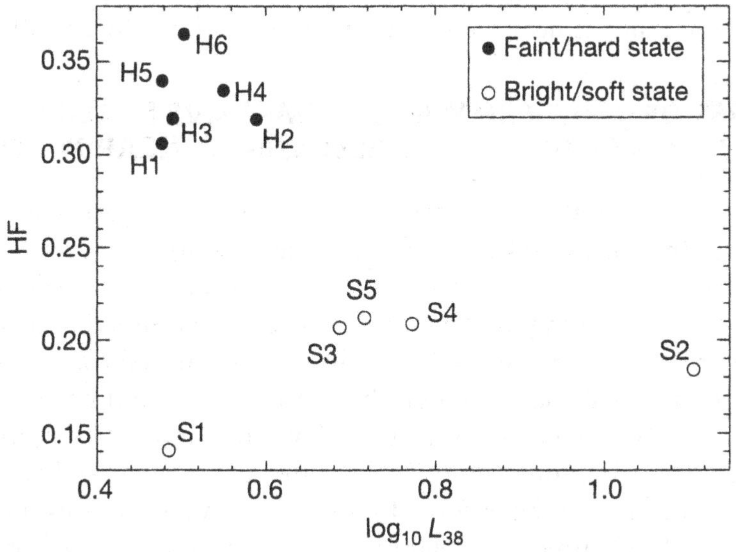

FIGURE 11.14 The X-ray luminosity and hard flux fraction (HF) for the 11 archival HETGS observations of GRS 1915+105. L_{38} is the X-ray luminosity in units of 10^{38} reg/s. HF is defined as the ratio of the unabsorbed continuum flux from 0.7–1.4 Å (8.6–18 keV) to 0.7–4.1 Å (3–18 keV). The 11 observations are classified as hard or soft based on previous X-ray studies (Belloni et al., 2000); as expected, the hard states have a higher HF (Neilsen and Lee, 2009).

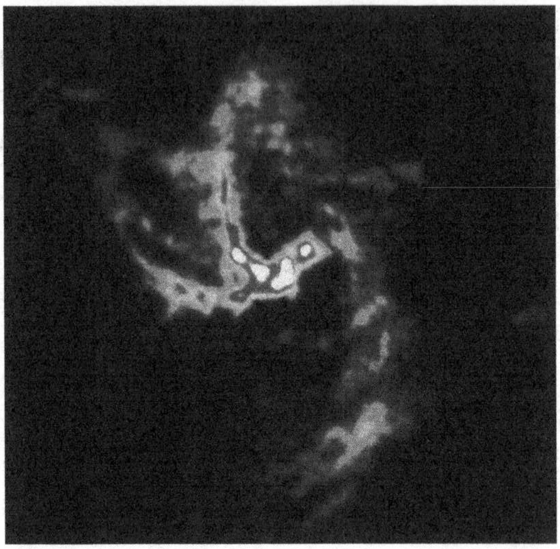

FIGURE 11.15 Gas clouds swirling around the black hole at the Milky Way's center as seen in radio emission (Reich, 2013).

neutron star called a *magnetar*, which induces various kinds of plasma activities in the Crab Nebula.

Figure 11.15 is the image of gas clouds swirling around the black hole at the Milky Way's center as seen in radio emission by the Very Large Array (VLA) of radio telescopes near Socorro in New Mexico. As it was assumed, the flare was coming from the activities of a magnetar in the gas clouds, and then, it may provide a useful tool to monitor the gas dynamics near the supermassive black hole (Reich, 2013).

11.5.3 BURST OF γ-RAY FROM A SUPERMASSIVE BLACK HOLE BREAKING APART AND SWALLOWING A NEARBY STAR

On 28 March 2011, the Swift team of astronomers was put on alert: a new γ-ray source had appeared in the northern sky. This was not big news for Swift, a satellite designed to look for γ-ray bursts (GRBs), which are a class of transient sources caused by the violent death of fast-spinning massive stars. However, continued observations of the event revealed that something entirely new had been detected (Burrows et al., 2011).

Whereas GRBs have a short-lived bright phase of γ-ray and hard X-ray emission lasting some ten seconds followed by a long (>1 month), smooth decay, the new source maintained a very high luminosity ($\sim 10^{27}$ kW) and a strong variability for more than a month. The peak value of the observed X-ray (1–10 keV) luminosity, in fact, reached $\sim 10^{48}$ erg/s, some 3×10^{14} times the solar luminosity. The total amount of X-ray energy emitted from this source in the period of more than a month was $\sim 2 \times 10^{53}$ erg; it will take more than a trillion years for the Sun to emit this amount of energy at the current luminosity. In addition, it is noteworthy that a wide range of spectrum in the electromagnetic radiation has been observed not only in γ-ray and X-ray but in optical (a few eV) and radio (~ 0.1 mm–10 m) ranges.

In addition, follow-up observations in the radio waveband showed that the new source was expanding with a velocity close to that of light (Zauderer et al., 2011). Finally, optical observation placed the transient at the center of a distant (~6 billion light years) galaxy (Levan et al., 2011).

The new source, called Swift J164449.3+573451, is now considered to be the electromagnetic signature of the tidal disruption of a star by the massive black hole sitting at the center of its host galaxy; it is a process that has been theoretically predicted earlier (Rees, 1988; Evans & Kochanek, 1989). We now analyze such a process in some detail.

Let us consider a situation in which a stellar object (called a star) with mass M_s and radius R_s is located at a distant R ($>R_B$: Schwarzschild radius) from a black hole with mass M_B. The star is a sphere consisting mostly of hydrogen plasma, bound by gravitational attraction, as with the Sun.

Now, the strength of the gravity is proportional to the mass and inversely proportional to the square of the distance. The side of the star facing the black hole is attracted to the black hole with the force $\sim M_B/(R-R_s)^2$; the other side is attracted to the black hole with the force $\sim M_B/(R+R_s)^2$. If the balance, which is the tidal force, sufficiently exceeds the gravitational binding force, $\sim M_s/R_s^2$, of the star, then the star is disrupted in plasma clouds. The resultant conditions are

$$R_B < R < R_s\left(\frac{M_B}{M_s}\right)^{1/3}. \tag{11.15}$$

Hence, when $M_B \simeq 10^6 M_S$ and $M_s \simeq M_S$, we numerically find

$$3\times10^6\ \text{km} < R < 7\times10^7\ \text{km}.$$

This is the condition for a star being disrupted, ground out, and turned into a huge chunk of plasmas.

The plasmas so created fall further into the gravitational region of the black hole and thereby form accretion disks in the way described in Sec. 11.4.3 and Sec. 11.5.1. They then plunge into super-strong gravitational fields near R_B (Schwarzschild radius), are heated to ultra-high temperatures, and thereby emit copious high-energy radiation such as γ-ray and X-ray. Furthermore, the accretion disks themselves may form large antennas for the emission of radio waves.

It would indeed be a phenomenal incident to conceive a huge black hole gulping star-sized plasmas, which abundantly emit the entire spectrum of electromagnetic radiation.

12

DAWN OF GRAVITATIONAL-WAVE ASTRONOMY

The 2017 Nobel Prize in Physics was awarded for the construction of, and the first direct detection of the gravitational waves by, the Laser Interferometer Gravitational-Wave Observatory (LIGO) operated jointly by the California Institute of Technology (Caltech) and the Massachusetts Institute of Technology (MIT) (Abbott et al., 2016a); Kip Thorne and Barry Barish of Caltech and Rainer Weiss of MIT received the Prize.

In 1916, Einstein predicted the existence of gravitational waves through the linearized weak-field solutions to the field equations of general relativity. The transverse waves of spatial variation in the gravitational field, generated by time variations of the mass quadrupole moment of the source, were predicted to travel at the speed of light (Einstein, 1916, 1918).

Gravitational-wave astronomy exploits multiple, widely separated detectors to distinguish gravitational waves from local instrumental and environmental noise, to provide source sky localization, and to measure wave polarization. The LIGO sites each operate a single Advanced LIGO detector (Aasi et al., 2015), a modified Michelson interferometer that measures gravitational-wave strain as a difference in length of its orthogonal arms. Each arm is formed by two mirrors, acting as test masses, separated by $L_x = L_y = L = 4$ km. A passing gravitational wave effectively alters the arm lengths such that the measured difference is $\Delta L(t) = \delta L_x - \delta L_y = h(t)L$, where $h(t)$ is the gravitational-wave strain amplitude projected onto the detector.

Theoretician Thorne and experimentalist Weiss originally conceived of the research project some 40 years ago; Barish then led this gigantic endeavor to success.

The achievements undoubtedly point to the dawn of gravitational-wave astronomy. The issues evidently involve the neutron stars and the black holes, which we studied closely in the preceding chapter. We begin this chapter with a discussion on the binary pulsars that played a cardinal role in the detection of the gravitational waves.

12.1 HULSE–TAYLOR BINARY PULSARS

From the beginning, it was recognized that gravitational-wave amplitudes would be remarkably small, because gravity is a weak force.

We here note, however, that neutron stars and black holes are unique astronomical objects that may exert exceedingly intense gravitational forces around themselves. The surface gravity of a neutron star with a solar mass and a radius of 10 km, for instance, is 1.4×10^{11} times as intense as the surface gravity on the earth. If such a neutron star performs an orbital motion, then the associated surface gravity may generate a significant amount of gravitational radiation, which will, in turn, affect the orbital behavior itself.

The discovery of the binary pulsar system PSR1913+16 at the Arecibo Observatory in Puerto Rico (Hulse & Taylor, 1975) and subsequent observations of pulse arrival times from this system between September 1974 and March 1981 (Taylor & Weisberg, 1982) were sufficient to yield a solution for the component masses and the absolute size of the orbit. The energy loss was exposed in the reduction of the orbital parameters, clearly demonstrating the existence of the gravitational waves.

The detailed features of a binary system may be decoded from observation of the pulse arrival times and the Doppler effect (cf. Sec. 11.3.1). The total mass is almost equally distributed between the pulsar (p) and its unseen companion (c), with $M_p = 1.42 \pm 0.06\ M_S$ and $M_c = 1.41 \pm 0.06\ M_S$. They perform orbital motion for a period of 7.75 hours along an elliptic trajectory with the mutual distance varying from 4.8 R_S to 1.1 R_S. On the basis of those data, we may estimate the rate at which the orbital period should decay as energy is lost from the system via gravitational radiation.

According to the general relativistic quadrupole formula, one should expect for the PSR1913+16 system an orbital period derivative $\dot{P}_b = (-2.403 \pm 0.005) \times 10^{-12}$. This means the orbital period decreases by 6.7×10^{-8} s in a period; revolution shrinks by 3.5 m in a year. Observations by Taylor and Weisberg yield the measured value $\dot{P}_b = (-2.30 \pm 0.22) \times 10^{-12}$. The excellent agreement provides compelling evidence for the existence of gravitational radiation. Figure 12.1 summarizes comparison between observations and general relativistic calculations.

This discovery, along with emerging astrophysical understanding (Press & Thorne, 1972), led to the recognition that direct observations of the amplitude and phase of gravitational waves would enable studies of additional relativistic systems and provide new tests of general relativity, especially in the dynamic strong-field regime.

12.2 GW150915: THE FIRST SIGNALS FOR LIGO

On September 14, 2015 at 09:50:45 UTC, the two modified Michelson interferometer detectors of LIGO, separated by 3000 km at Hanford, Washington and at Livingston,

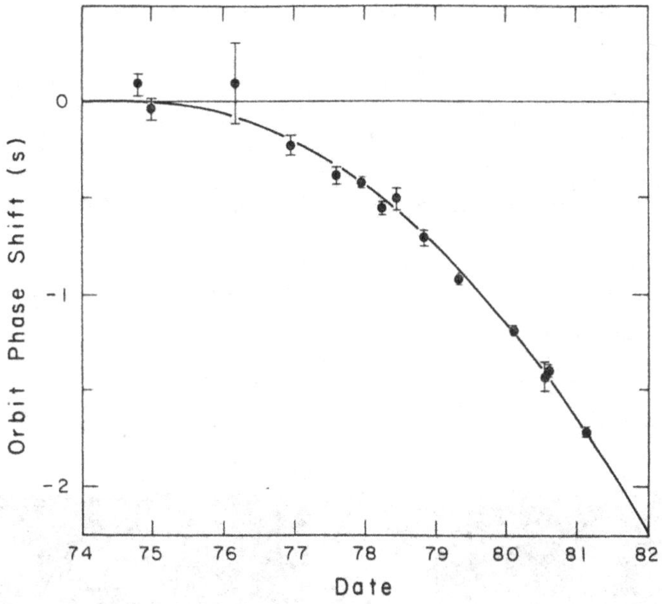

FIGURE 12.1 Emission of gravitational waves deduced from the orbital phase residuals on PSR1913+16 observed between September 1974 and March 1981. The points are the observed residuals; the curvature of the parabola corresponds to the general relativistic prediction: $\dot{P}_b = -2.40 \times 10^{-12}$; if there is no gravitational radiation, a straight line may result (Taylor & Weisberg, 1982).

Louisiana, simultaneously observed transient gravitational-wave signals, shown in Figure 12.2 (Abbott et al., 2016a).

To many, the timing of the signal seemed too good to be true, for it was only several days previously that the collaboration had completed a five-year upgrade to its instruments. Moreover, the LIGO collaboration had also given a small number of its members the power to inject fake signals and to hide whether they were real or simulated in order to test the team's responses. After a long day of calls and e-mails, it was determined that no such "blind injection" had occurred (Castelvecchi, 2016).

The team then decided to take data for another month before beginning a full analysis: the researchers needed to record the natural noise present in their detectors to have something to compare with the signal. They concluded that the odds of noise producing that loud pattern—and the very same pattern in both Louisiana and Washington at about the same time—were so low that it should only occur by chance less than once every 203,000 years (Castelvecchi, 2016).

12.2.1 INFORMATION EXTRACTED FROM THE SIGNALS

The basic features of GW150914 point to it being produced by the coalescence of two black holes—i.e., their orbital inspiral, merger, and subsequent final black hole ringdown. Over 0.2 s, the signal increases in frequency and amplitude in about 8 cycles

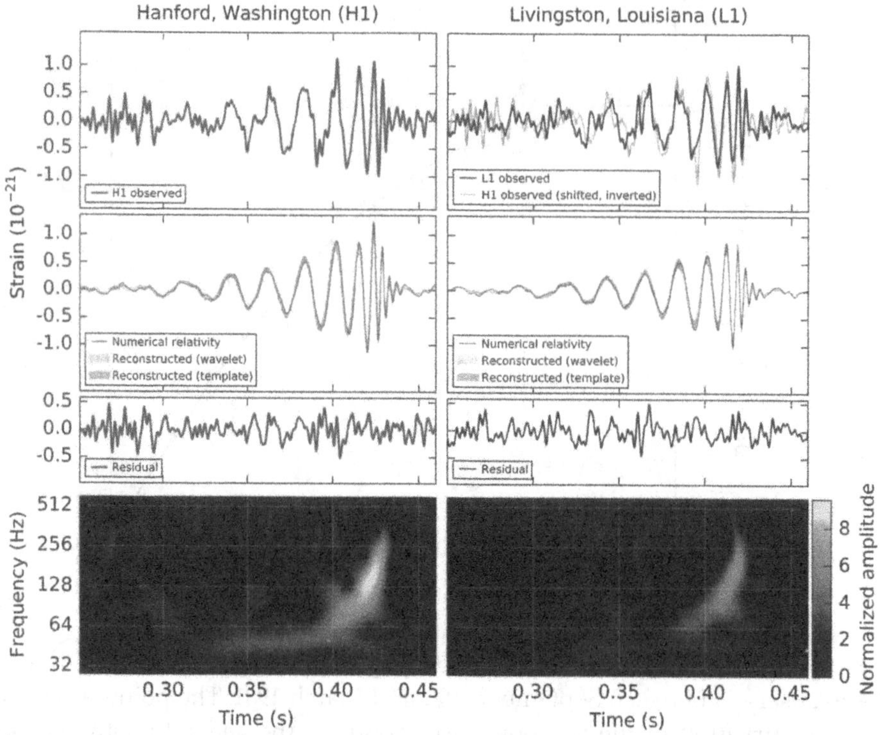

FIGURE 12.2 The gravitational-wave event GW150914 observed by the LIGO Hanford (H1, left column panels) and Livingston (L1. right column panels) detectors. *Top row, left:* H1 strain. *Top row, right:* L1 strain. GW150914 arrived first at L1 and 6.9 ms later at H1; for visual comparison, the H1 data are also shown, shifted in time by this amount and inverted (to account for the detectors' relative orientations). *Second row:* Gravitational-wave strain projected onto each detector in the 35–350 Hz band. Solid lines show a numerical relativity waveform for a system with parameters consistent with those recovered from GW150914 (Mroué et al., 2013) confirmed to 99.9% by an independent calculation based on Campanelli et al. (2006). *Third row:* Residuals after subtracting the filtered numerical relativity waveform from the filtered detector time series. *Bottom row:* A time-frequency representation (Chatterji et al., 2004) of the strain data, showing the signal frequency increasing over time (Abbott et al., 2016a).

from 35 to 150 Hz, where the amplitude reaches a maximum. The most plausible explanation for this evolution is the inspiral of two orbiting masses m_1 and m_2 due to gravitational-wave emission (cf. Sec. 12.1). At the lower frequencies, such evolution is characterized by the chirp mass (Blanchet et al., 1995)

$$M_{\text{chirp}} = \frac{(m_1 m_2)^{3/5}}{(m_1 + m_2)^{1/5}} = \frac{c^3}{G}\left[\frac{5}{96}\pi^{-8/3} f^{-11/3} \dot{f}\right]^{3/5}, \tag{12.1}$$

where f and $\dot{f}$ are the observed frequency and its time derivative and G and c are the gravitational constant and speed of light. Estimating f and $\dot{f}$ from the data in Figure 12.2, we obtain a chirp mass $\sim 30M_S$, implying that the total mass $M = m_1 + m_2$

is $\gtrsim$70 M_S in the detector frame. This bounds the sum of the Schwarzschild radii of the binary components to $2GM/c^2 \gtrsim 210$ km. To reach an orbital frequency of 75 Hz (half the gravitational-wave frequency) the objects must have been very close and very compact; equal Newtonian point masses orbiting at this frequency would be only $\simeq$350 km apart.

From the waveforms, the researchers were thus able to deduce that one black hole was about 36 times the mass of the Sun, and the other was about 29 times the solar mass. Those then coalesce into a black hole of 62 solar masses, with a mass defect of three solar masses. The estimated total energy radiated in gravitational waves is thus $\sim 3.0 M_S c^2$, which traveled across the Universe for 1.3 billion years of luminosity distance to reach us. A peak gravitational-wave luminosity of $\sim 3.6 \times 10^{56}$ erg/s, equivalent to $\sim 200 M_S c^2$/s, may have been reached.

12.2.2 ITEMS TO BE ENSURED WITH THE SIGNALS

Undoubtedly, the first direct detection of the gravitational waves exhibited in Figure 12.2 is quite convincing. There seem to remain some items to be ensured in the interpretation, however.

1. In a number of microquasars hitherto observed (cf. Sec. 11.5.1), the masses of the stellar black holes definitely established have not exceeded $15M_S$. Analysis of GW150914 has doubled this record at a stroke (i.e., $36M_S$ and $29M_S$), and then doubled it again (i.e., $62M_S$). Neither have we seen a binary of two black holes as yet. It therefore seems premature to accept the source parameters including the black hole merger, derived from comparison of post-Newtonian general-relativistic simulations with only the 0.2 s waveform; separate, independent investigations may be needed for reconfirmation. We may recall in these connections, three analogous signals have been obtained (Abbott et al., 2016b, 2017a, b); further studies inclusive of these signals should be illuminating.
2. A connection with the electromagnetic-wave—γ-ray, X-ray, visible light, infrared, radio wave—astronomy needs to be established, especially in light of the Swift J164449.3+573451 observation of a wide range of spectrum in the electromagnetic radiation, with the peak luminosity in excess of $\sim 10^{48}$ erg/s, some 3×10^{14} times the solar luminosity, as described in Sec. 11.5.3.

12.3 OBSERVATION OF COLLIDING BINARY NEUTRON STARS

"Sometimes nature can be generous"—a *Nature* article begins by so quipping (Miller, 2017). Its generosity was on full display on 17 August, 2017, when two neutron stars with masses in the range 1.2–1.6 M_S spiraled together some 130 million light years away. The event, called GW170817 (Abbott et al., 2017c), may provide even greater treasure than the black hole mergers presented in the preceding section, because it produced both gravitational waves and electromagnetic radiation. Figure 12.3 displays an illustrative drawing for the merger of binary neutron-star system.

GW170817 was observed by the Advanced LIGO as well as by the Advanced Virgo gravitational-wave detector, situated outside Pisa in Italy, and its distance from the US-based LIGO detectors allowed the location of GW170817 on the sky to be determined with an uncertainty of about 30 square degrees, compared with 600 square degrees or more for the detections by the LIGO detectors alone.

GW170817 was observed also in the γ-ray burst GRB170817A, detected by Fermi-GBM (gamma-ray burst monitor) 1.7 s after the coalescence (Connaughton, 2017). This observation corroborates the hypothesis of a neutron star merger and the first direct evidence of a link between these mergers and short γ-ray bursts.

GW170817 was observed further in X-ray, visible light, and infrared (Arcavi et al., 2017; Pian et al., 2017; Troja et al., 2017; Smartt et al., 2017; Kasen et al., 2017). As a result, the event provides tests of alternative theories of gravity. It also delivers strong evidence for the formation path of at least some of the heavy elements in the Universe (those much heavier than iron).

Since the Hulse–Taylor discovery, radio pulsar surveys have found several more binary neutron-star (BNS) systems in our Galaxy (Manchester et al., 2005). Understanding the orbital dynamics of these systems inspired detailed theoretical

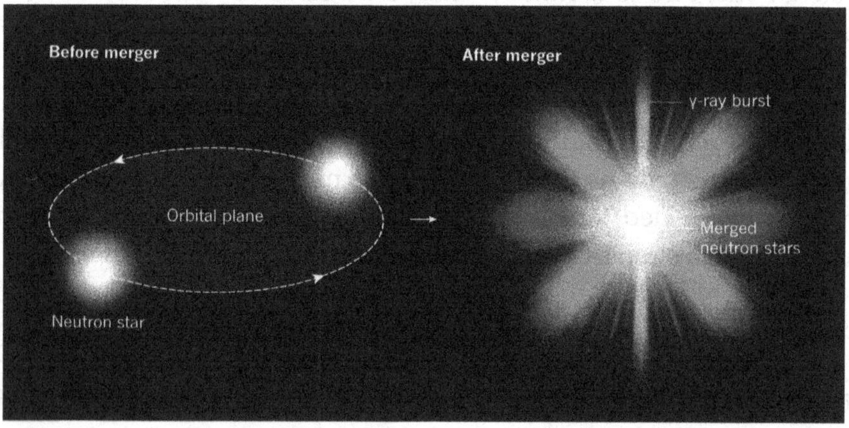

FIGURE 12.3 Illustrative drawing for the merger of binary neutron-star system (Miller, 2017). Gravitational waves have been detected from the coalescence of two orbiting neutron stars (Abbott et al., 2017c). Unlike the previous discoveries of the gravitational waves (Abbott et al., 2016a, b, 2017a, b), the event has been observed across the electromagnetic spectrum. The Fermi Gamma-ray Space Telescope saw a flush of g-rays just two seconds after the neutron-star merger (Connaughton, 2017). The flash is consistent with a cosmic explosion called a g-ray burst. In addition, five papers (Arcavi et al., 2017; Pian et al., 2017; Troja et al., 2017; Smartt et al., 2017; Kasen et al., 2017) report the emission of X-rays, optical light (blue), and infrared light (red) from the merged neutron stars.

predictions for gravitational-wave signals from compact binaries (Blanchet et al., 1995, 2004). Models of the population of compact binaries, informed by the known binary pulsars, predict that the network of advanced gravitational-wave detectors operating at design sensitivity may possibly observe between one BNS merger every few years to hundreds per year (Abbott et al., 2017c).

For all of these reasons, GW170817 represents a remarkable opportunity to make major progress in multiple fields of physics and astronomy, and it whets our appetite for the many expected observations of BNS mergers in future campaigns.

APPENDIX I: THE δ-FUNCTIONS

The function $\delta(x)$ satisfying

$$\delta(x) = \begin{cases} 0 & (x \neq 0) \\ \infty & (x = 0) \end{cases} \tag{AI.1}$$

$$\int_{-\infty}^{\infty} dx\delta(x) = 1 \tag{AI.2}$$

is called the delta function. For an arbitrary function $f(x)$ defined over the domain $-\infty < x < \infty$, the following relations hold:

$$\int_{-\infty}^{\infty} dx f(x)\delta(x) = f(0), \tag{AI.3}$$

$$\int_{a}^{b} dx f(x)\delta(x - x_0) = f(x_0) \quad (a < x_0 < b). \tag{AI.4}$$

Various limiting forms exist as analytic expressions for the delta function. For example,

$$\delta(x) = \lim_{\eta \to 0} \frac{1}{\eta\sqrt{\pi}} \exp\left(-\frac{x^2}{\eta^2}\right), \tag{AI.5}$$

$$\delta(x) = \lim_{\eta \to 0} \frac{\eta}{\pi(x^2 + \eta^2)} = -\lim_{\eta \to 0} \mathrm{Im} \frac{1}{\pi(x + i\eta)}, \tag{AI.6}$$

satisfy the foregoing relations ("Im" means the imaginary part).

A three-dimensional delta function $\delta(\mathbf{r})$ with respect to the spatial coordinates $\mathbf{r} = (x, y, z)$ can be produced from the product of the one-dimensional delta functions as

$$\delta(\mathbf{r}) = \delta(x)\delta(y)\delta(z). \tag{AI.7}$$

APPENDIX II: FOURIER ANALYSES AND APPLICATION

In this volume, we frequently use Fourier analyses of the physical quantities to describe such effects as fluctuations. This appendix summarizes the essence and application of the Fourier analyses.

Let a quantity, $a(\mathbf{r}, t)$, be a function of the space-time coordinates, $\mathbf{r}$ and t, satisfying the periodic boundary conditions with the volume $V = L^3$:

$$a(\mathbf{r},t) = a(\mathbf{r} + n_x L\mathbf{x} + n_y L\mathbf{y} + n_z L\mathbf{z}, t). \tag{AII.1}$$

Here, $\mathbf{x}$, $\mathbf{y}$, and $\mathbf{z}$ are the unit vectors in the x, y, and z directions; n_x, n_y, and n_z represent positive and negative integers including 0. In the text, we may sometimes consider the cases with $L = 1$; such a case is then referred to as a Fourier analysis with the periodic boundary conditions in a unit volume.

The space-time Fourier components of $a(\mathbf{r},t)$ are then calculated as

$$a_{\mathbf{k}}(\omega) = \int_{L^3} d\mathbf{r} \int_{-\infty}^{\infty} dt a(\mathbf{r},t) \exp[-i(\mathbf{k}\cdot\mathbf{r} - \omega t)]. \tag{AII.2}$$

Since we have adopted the periodic boundary conditions for a cube of volume L^3, the wave vector $\mathbf{k}$ takes on discrete values,

$$\mathbf{k} = \frac{2\pi}{L}\left(n_x, n_y, n_z\right). \tag{AII.3}$$

The original function $a(\mathbf{r},t)$ may be reproduced from (AII.2) through the inverse Fourier transformation as

$$a(\mathbf{r},t) = \frac{1}{L^3}\sum_{\mathbf{k}} \int_{-\infty}^{\infty} \frac{d\omega}{2\pi} a_{\mathbf{k}}(\omega) \exp\left[i(\mathbf{k}\cdot\mathbf{r} - \omega t)\right], \tag{AII.4}$$

where the summation with respect to $\mathbf{k}$ is performed over those discrete values specified by (AII.3).

As (AII.3) indicates, a volume $(2\pi/L)^3$ in the three-dimensional wave-number space (k_x, k_y, k_z) corresponds to a wave vector pertaining to the summation of (AII.4). In the limit of $L \to \infty$, the summation in (AII.4) turns into a three-dimensional integration with respect to $\mathbf{k}$. In this case, since

$$\sum_{\mathbf{k}} \to L^3 \int_{-\infty}^{\infty} \frac{dk_x}{2\pi} \int_{-\infty}^{\infty} \frac{dk_y}{2\pi} \int_{-\infty}^{\infty} \frac{dk_z}{2\pi} \equiv L^3 \int_{-\infty}^{\infty} \frac{d\mathbf{k}}{(2\pi)^3}, \tag{AII.5}$$

(AII.4) is transformed as

$$a(\mathbf{r},t)=\int_{-\infty}^{\infty}\frac{d\mathbf{k}}{(2\pi)^3}\int_{-\infty}^{\infty}\frac{d\omega}{2\pi}a_{\mathbf{k}}(\omega)\exp[i(\mathbf{k}\cdot\mathbf{r}-\omega t)]. \tag{AII.6}$$

The three-dimensional delta function $\delta(\mathbf{r})$ introduced in (AI.7) of Appendix I is Fourier transformed as

$$\int d\mathbf{r}\delta(\mathbf{r})\exp(-i\mathbf{k}\cdot\mathbf{r})=1. \tag{AII.7}$$

Hence, we find

$$\delta(\mathbf{r})=\int_{-\infty}^{\infty}\frac{d\mathbf{k}}{(2\pi)^3}\exp(i\mathbf{k}\cdot\mathbf{r}) \tag{AII.8}$$

with the aid of (AII.6).

As an application of the Fourier analysis, we take up the problem of solving (1.12) in Sec. 1.3.2. First, we multiply both sides of (1.12) by $\exp(-i\mathbf{k}\cdot\mathbf{r})$ and integrate the result with respect to $\mathbf{r}$. Assuming that $\phi(\mathbf{r})$ approaches zero sufficiently fast in the limit $r\to\infty$, we obtain by partial integrations

$$\begin{aligned}\int d\mathbf{r}\exp(-i\mathbf{k}\cdot\mathbf{r})\nabla^2\phi(\mathbf{r}) &= \int d\mathbf{r}(i\mathbf{k})\exp(-i\mathbf{k}\cdot\mathbf{r})\cdot\nabla\phi(\mathbf{r})\\ &= -k^2\int d\mathbf{r}\exp(-i\mathbf{k}\cdot\mathbf{r})\cdot\phi(\mathbf{r})\\ &= -k^2\phi_{\mathbf{k}}\end{aligned} \tag{AII.9}$$

That the aforementioned assumption is satisfied can be confirmed a posteriori. The final line of (AII.9) is the definition of $\phi_{\mathbf{k}}$.

With the aid of (AII.7) and (AII.9), we find that (1.12) reduces to

$$\phi_{\mathbf{k}}=\frac{4\pi Z_0 e}{k^2+\lambda_D^{-2}}. \tag{AII.10}$$

Carrying out the inverse Fourier transformation of (AII.10) in accord with (AII.6), where θ refers to the angle between $\mathbf{k}$ and $\mathbf{r}$, we obtain

$$\begin{aligned}\phi(\mathbf{r}) &= \frac{1}{(2\pi)^3}\int_0^{n}2\pi\sin\theta d\theta\int_0^{\infty}k^2dk\frac{4\pi Z_0 e}{k^2+\lambda_D^{-2}}\exp(-ikr\cos\theta)\\ &= \frac{2Z_0 e}{\pi r}\int_0^{\infty}dk\frac{k}{k^2+\lambda_D^{-2}}\sin(kr).\end{aligned}$$

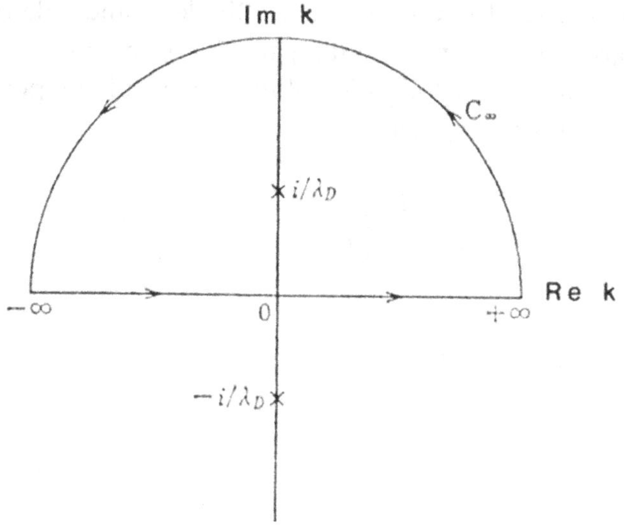

FIGURE AII.1 Contour of integration in Eq. (AII.12).

We note that the integrand is an even function of k and that sin(kr) is the imaginary part (Im) of exp(ikr) = cos(kr) + $i$sin(kr). Hence, we may rewrite the equation above in the form,

$$\phi(\mathbf{r}) = \frac{Z_0 e}{\pi r} \mathrm{Im} \int_{-\infty}^{\infty} dk \frac{k}{k^2 + \lambda_D^{-2}} \exp(ikr). \tag{AII.11}$$

To carry out this integral, we close the contour of integration by an infinite semicircle C_∞ on the upper half of the complex k plane as shown in Figure AII.1. The integrand of (AII.11) has the first-order poles at $k = \pm i/\lambda_D$; the residue at $k = i/\lambda_D$ enclosed by the contour of integration is

$$\lim_{k \to i/\lambda_D} \frac{(k - i/\lambda_D)}{k^2 + \lambda_D^{-2}} \exp(ikr) = \frac{1}{2} \exp\left(-\frac{r}{\lambda_D}\right).$$

In light of Cauchy's theorem, we obtain

$$\int_{-\infty}^{\infty} dk \frac{k}{k^2 + \lambda_D^{-2}} \exp(ikr) + \int_{C_\infty} dk \frac{k}{k^2 + \lambda_D^{-2}} \exp(ikr) = i\pi \exp\left(-\frac{r}{\lambda_D}\right). \tag{AII.12}$$

The second term on the left-hand side of (AII.12) is the integration along the infinite semicircle on the upper-half plane shown in Figure AII.1. We transform it into an integration from 0 to π with respect to θ by putting $k = K\exp(i\theta)$; in the limit of $K \to \infty$, this term vanishes. Substitution of (AII.12) to (AII.11) yields (AII.9).

If, in particular, we let the second term on the left-hand side of (1.12) approach 0, i.e., $\lambda_D^{-2} \to 0$, $\phi(\mathbf{r})$ should coincide with $\phi_0(\mathbf{r}) = Z_0 e/r$, the bare Coulomb potential. Consequently, the Fourier transformation of the bare Coulomb potential is given by (AII.10) where $\lambda_D^{-2} \to 0$ is taken, that is,

$$\frac{Ze}{r} = \frac{1}{L^3} \sum_{\mathbf{k}} \frac{4\pi Ze}{k^2} \exp(i\mathbf{k} \cdot \mathbf{r}). \tag{AII.13}$$

APPENDIX III: THE FLUCTUATION-DISSIPATION THEOREM

In the theory of the many-particle system in thermodynamic equilibrium, the fluctuation–dissipation theorem provides a rigorous connection between the spectral functions and the imaginary parts of the relevant linear response functions (Callen & Welton, 1951; Kubo, 1957). The theorem relates the canonically (or grand canonically) averaged commutator [,] and anticommutator { , } of any pair of Hermitian operators, such as the number densities evaluated at two different points in space and time. The average of such a commutator is related to a response function, while the average of an anticommutator gives a correlation function, which turns into a structure factor or a spectral function of fluctuations after Fourier transformations. It may therefore be said that the theorem possesses a form unique in physics, relating the properties of the system in equilibrium (i.e., fluctuations) with the parameters that characterize the irreversible processes, that is, the imaginary parts of the response functions. The contents of the fluctuation–dissipation theorem are summarized in this appendix.

We begin with a set of external disturbances,

$$a(\mathbf{r},t) = a\exp\left[i(\mathbf{k}\cdot\mathbf{r} - \omega t) + 0t\right] + \text{cc}, \tag{AIII.1}$$

applied to a many-particle system in thermodynamic equilibrium; the system would be uniform in the absence of the disturbances (cc stands for the complex conjugate and 0 designates a positive infinitesimal).

These disturbances then produce an external Hamiltonian,

$$H_{\text{ext}}(t) = -\sum\int_V d\mathbf{r}A(\mathbf{r})a(\mathbf{r},t), \tag{AIII.2}$$

where A represents the physical quantity observable in the system that is coupled to the disturbance (AIII.1); the sum in (AIII.2) goes over the set of the disturbances (AIII.1). The total Hamiltonian written as the sum of the unperturbed and external contributions,

$$H_{\text{tot}} = H + H_{\text{ext}}(t), \tag{AIII.3}$$

then drives the system out of equilibrium according to the Heisenberg equation of motion (see 2.20).

A physical quantity B, which can be the same as A, of the system is perturbed and thereby deviates from its average value by $\delta B(\mathbf{r}, t)$. Within the framework of the linear response formalism (e.g., Sec. 2.2.1), the deviations may be expressed as

$$\delta B(\mathbf{r},t) = B\exp[i(\mathbf{k}\cdot\mathbf{r}-\omega t)+0t]+\text{cc}. \quad \text{(AIII.4)}$$

In this case,

$$\chi_{BA}(\mathbf{k},\omega) = B/a \quad \text{(AIII.5)}$$

gives a linear response function in its general form. Explicit calculations with the Hamiltonian (AIII.3) yield a commutator expression for the response function:

$$\chi_{BA}(\mathbf{k},\omega) = \frac{i}{\hbar}\int d\mathbf{r}\int_0^{\infty} dt\left\langle\left[B(\mathbf{r}'+\mathbf{r},t'+t),A(\mathbf{r}',t')\right]\right\rangle\exp[-i(\mathbf{k}\cdot\mathbf{r}-\omega t)]. \quad \text{(AIII.6)}$$

Here, $A(\mathbf{r}, t)$ and $B(\mathbf{r}, t)$ are the Heisenberg operators evolving with the unperturbed Hamiltonian as in (2.4) and $\langle\cdots\rangle$ refers to the expectation value in the unperturbed equilibrium state. The linear response functions therefore depend only on the system properties without perturbations.

The physical quantities A and B fluctuate in space and time even in a system under thermodynamic equilibrium. The correlation function between them is defined in terms of the statistical average of the anticommutator as

$$C_{BA}(\mathbf{r},t) = \frac{1}{2}\left\langle\left\{B(\mathbf{r}'+\mathbf{r},t'+t),A(\mathbf{r}',t')\right\}\right\rangle. \quad \text{(AIII.7)}$$

Fourier transformation of such a correlation function yields a structure factor or a spectral function of the fluctuations:

$$S_{BA}(\mathbf{k},\omega) = \frac{1}{2\pi}\int d\mathbf{r}\int_{-\infty}^{\infty} dt C_{BA}(\mathbf{r},t)\exp[-i(\mathbf{k}\cdot\mathbf{r}-\omega t)]. \quad \text{(AIII.8)}$$

It is then connected with the linear response function (AIII.6) via the *fluctuation–dissipation theorem*,

$$S_{BA}(\mathbf{k},\omega) = -\frac{i\hbar}{4\pi}\coth\left(\frac{\hbar\omega}{2k_BT}\right)\left[\chi_{BA}(\mathbf{k},\omega)-\chi_{AB}(-\mathbf{k},-\omega)\right]. \quad \text{(AIII.9)}$$

In conjunction with the considerations related to static responses in many-particle systems, it is instructive to treat a specific case of external disturbance, which in terms of the spatial Fourier component takes the form

$$\frac{\partial B}{\partial a} \equiv \lim_{\mathbf{k}\to 0}\frac{\delta B(\mathbf{k},t=0)}{a(\mathbf{k})} = \lim_{\mathbf{k}\to 0}\frac{1}{\pi}\int_{-\infty}^{\infty} d\omega\frac{\chi''_{BA}(\mathbf{k},\omega)}{\omega}. \quad \text{(AIII.10)}$$

The Fourier component $\delta B(\mathbf{k}, t=0)$ of the induced fluctuation is then calculated in terms of the *relaxation functions*

$$\chi''_{BA}(\mathbf{k},\omega) = \frac{1}{2\hbar}\int d\mathbf{r}\int_{-\infty}^{\infty} dt \langle [B(\mathbf{r}'+\mathbf{r},t'+t), A(\mathbf{r}',t')] \rangle \exp[-i(k\cdot r - \omega t)] \quad \text{(AIII.11)}$$

as

$$\frac{\delta B(\mathbf{k},t=0)}{a(\mathbf{k})} = \frac{1}{\pi}\int_{-\infty}^{\infty} d\omega \frac{\chi''_{BA}(\mathbf{k},\omega)}{\omega} = \frac{1}{2k_B T}\langle [B(\mathbf{k},t=0), A(-\mathbf{k},t=0)] \rangle. \quad \text{(AIII.12)}$$

Thermodynamic sum rules for the relaxation functions are finally obtained in the long wavelength limit of (AIII.12) as

$$\frac{\partial B}{\partial a} \equiv \lim_{\mathbf{k}\to 0} \frac{\delta B(\mathbf{k},t=0)}{a(\mathbf{k})} = \lim_{\mathbf{k}\to 0} \frac{1}{\pi}\int_{-\infty}^{\infty} d\omega \frac{\chi''_{BA}(\mathbf{k},\omega)}{\omega}. \quad \text{(AIII.13)}$$

The compressibility sum rule and the spin-susceptibility sum rule in Sec. 2.2.6 are typical examples of these thermodynamic sum rules.

APPENDIX IV: FERMI INTEGRALS

In the treatment of a free-electron system at finite temperatures, it is useful to define the Fermi integrals:

$$I_\nu(\alpha) \equiv \int_0^\infty dx \frac{x^\nu}{\exp(x-\alpha)+1}. \tag{AIV.1}$$

For the electrons in the paramagnetic state (2.51), the normalization condition (2.39) is then expressed as

$$I_{1/2}(\alpha) = \frac{2}{3}\theta^{-3/2}, \tag{AIV.2}$$

where $\alpha = \beta\mu_\sigma$, β ($=1/k_BT$) is the inverse temperature in energy unit, and θ represents the ratio between that temperature and the Fermi energy, E_F, as expressed by (1.5).

The Fermi pressure p_0 of the free-electron system is likewise expressed as

$$\beta p_0 = n\theta^{3/2} I_{3/2}(\alpha). \tag{AIV.3}$$

The ideal-gas contribution to the free energy per unit volume is then calculated as

$$\beta f_0 = n\alpha - \beta p_0. \tag{AIV.4}$$

Useful fitting formulas for the chemical potential and the Fermi pressure are

$$\beta\mu_\sigma = -\frac{3}{2}\ln\theta + \ln\frac{4}{3\sqrt{\pi}} + \frac{A\theta^{-(b+1)} + B\theta^{-(b+1)/2}}{1 + A\theta^{-b}} \tag{AIV.5}$$

with $A = 0.25954$, $B = 0.072$, and $b = 0.858$; and

$$\frac{\beta p_0}{n} = 1 + \frac{2}{5}\frac{X\theta^{-(y+1)} + Y\theta^{-(y+1)/2}}{1 + X\theta^{-y}}, \tag{AIV.6}$$

with $X = 0.27232$, $Y = 0.145$, and $y = 1.044$. Maximum deviations of (AIV.5) from the exact values determined from (AIV.2) are about 0.26% at $\theta \sim 0.05$; those of (AIV.6) from the exact values determined from (AIV.3) are about 0.26% at $\theta \sim 5$.

In the classical limit—when $\theta \gg 1$—the Fermi integrals may be expanded as (e.g., Pathria, 1972)

$$I_\nu(\alpha) = \Gamma(\nu+1)\sum_{s=1}^{\infty}(-1)^{s+1}\exp(s\alpha)s^{-(\nu+1)}, \tag{AIV.7}$$

where

$$\Gamma(z) = \int_0^\infty dt\, t^{z-1} \exp(-t) \quad (\text{Re}\, z > 0). \tag{AIV.8}$$

is the gamma function. In this limit, one thus has

$$\alpha = -\frac{3}{2}\ln\theta + \ln\frac{4}{3\sqrt{\pi}}, \tag{AIV.9}$$

$$\frac{\beta p_0}{n} = 1. \tag{AIV.10}$$

In the quantum limit of strong degeneracy—when $\theta \ll 1$—

$$I_\nu(\alpha) = \frac{\alpha^{\nu+1}}{\nu+1}\left[1 + \sum_{s=1}^{\infty} 2(1 - 2^{1-2s})\varsigma(2s)\frac{(\nu+1)!}{(\nu+1-2s)!}\alpha^{-2s}\right] \tag{AIV.11}$$

(Landau & Lifshitz, 1969), where

$$\varsigma(x) = \sum_{\nu-1}^{\infty} \frac{1}{\nu^x} \quad x > 1, \tag{AIV.12}$$

is Riemann's zeta function. Hence,

$$\mu_\sigma = E_F, \tag{AIV.13}$$

$$p_0 = (2/5) n E_F, \tag{AIV.14}$$

in this limit. For reference, we list some values of those functions:

$$\varsigma\left(\frac{3}{2}\right) = 2.612, \quad \varsigma\left(\frac{5}{2}\right) = 1.341, \quad \varsigma(3) = 1.202,$$

$$\varsigma(5) = 1.037, \quad \Gamma\left(\frac{3}{2}\right) = \frac{\sqrt{\pi}}{2}, \quad \Gamma\left(\frac{5}{2}\right) = \frac{3\sqrt{\pi}}{4}.$$

APPENDIX V: FUNCTIONAL DERIVATIVES

The functional derivative technique is closely related to the dielectric formulations and the density-functional theory treated in Chap. 2. In this appendix, fundamental relations in the functional derivatives are summarized.

Let $\mathbf{F}[f(x)]$ be a functional of a function $f(x)$ with a variable x defined over a domain, $a \le x \le b$. The functional derivative, $\delta\mathbf{F}/\delta f(x)$, is given in terms of the increment $\delta\mathbf{F}$ produced by an infinitesimal variation $\delta f(x)$ of $f(x)$ as

$$\delta\mathbf{F} = \int_a^b dx \frac{\delta\mathbf{F}}{\delta f(x)} \delta f(x). \tag{AV.1}$$

It has the following properties:

1. ***Identity relation:***

$$\frac{\delta f(x')}{\delta f(x)} = \delta(x - x'). \tag{AV.2}$$

2. ***Product rule:*** When a functional is expressed by a simple product of the functions $f(x_i)$ as

$$\mathbf{F} = \int_a^b dx_1 \cdots dx_N \prod_{i=1}^{N} f(x_i), \tag{AV.3}$$

then

$$\frac{\delta\mathbf{F}}{\delta f(x)} = N\int_a^b dx_2 \cdots dx_N \prod_{i=2}^{N} f(x_i). \tag{AV.4}$$

3. ***Chain rule:*** When a functional $\mathbf{F}$ is a functional of $\mathbf{G}(x')$ which in turn is a functional of $f(x)$, then

$$\frac{\delta\mathbf{F}}{\delta f(x)} = \int_a^b dx' \frac{\delta\mathbf{F}}{\delta\mathbf{G}(x')} \frac{\delta\mathbf{G}(x')}{\delta f(x)}. \tag{AV.5}$$

4. When a functional is given by

$$\mathbf{F} = \int_a^b dx F(f(x)), \tag{AV.6}$$

where $F(f(x))$ is a function of $f(x)$, then

$$\frac{\delta \mathbf{F}}{\delta f(x)} = \frac{dF(f(x))}{df(x)}. \tag{AV.7}$$

REFERENCES

Aasi, J. et al., 2015, *Class. Quant. Grav.* **32**, 074001.

Abbott, B. P. et al., 2016a, *Phys. Rev. Lett.* **116**, 061102.

Abbott, B. P. et al., 2016b, *Phys. Rev. Lett.* **116**, 241103.

Abbott, B. P. et al., 2017a, *Phys. Rev. Lett.* **118**, 221101.

Abbott, B. P. et al., 2017b, *Phys. Rev. Lett.* **119**, 141101.

Abbott, B. P. et al., 2017c, *Phys. Rev. Lett.* **119**, 161101.

Abe, R., 1959, *Prog. Theor. Phys.* **22**, 213.

Adli, E. et al., 2018, *Nature* **561**, 363.

Adriani, A. et al., 2018, *Nature* **555**, 216.

Alder, B. J. & T. E. Wainwright, 1959, *J. Chem. Phys.* **31**, 459.

Andersen, H. C., 1980, *J. Chem. Phys.* **72**, 2384.

Alastuey, A. & B. Jancovici, 1978, *Astrophys. J.* **226**, 1034.

Arcavi, I. et al., 2017, *Nature* **551**, 64.

Arnett, W. D. & J. W. Truran, 1969, *Astrophys. J.* **157**, 339.

Ashcroft, N. W. & N. D. Mermin, 1976, *Solid State Physics* (Sounders College Pub., Philadelphia, PA).

Aumann, H. H., C. M. Gillespie, Jr., & F. J. Low, 1969, *Astrophys. J. Lett.* **157**, L69.

Bahcall, J. N. & M. H. Pinsonneault, 1995, *Rev. Mod. Phys.* **67**, 781.

Bahcall, J. N. & R. K. Ulrich, 1988, *Rev. Mod. Phys.* **60**, 297.

Bahcall, J. N., W. F. Huebner, S. H. Lubow, P. D. Parker, & R. K. Ulrich, 1982, *Rev. Mod. Phys.* **54**, 767.

Barnes, C. A., 1971, *Adv. Nucl. Phys.* **4**, 133.

Baym, G., 1995, in *Elementary Processes in Dense Plasmas: Proc. Oji International Seminar*, edited by S. Ichimaru & S. Ogata (Addison-Wesley, Reading, MA), p. 3.

Baym, G. & C. J. Pethick, 1975, *Ann. Rev. Nucl. Sci.* **25**, 27.

Baym, G., C. J. Pethick, & P. Sutherland, 1971, *Astrophys. J.* **171**, 651.

Belloni, T., M. Klein-Wolt, M. Mendez, M. van der Klis, & J. van Paradijis, 2000, *Astron. Astrophys.* **355**, 271.

Binder, K., ed., 1979, *The Monte Carlo Method in Statistical Physics* (Springer, Berlin, Germany).

Binder, K., ed., 1992, *The Monte Carlo Method in Condensed Matter Physics* (Springer, Berlin, Germany).

Blanchet, L., T. Damour, B.R. Iyer, C.M. Will, & A.G. Wiseman, 1995, *Phys. Rev. Lett.* **74**, 3515.

Blanchet, L. T. Damour, B.R. Iyer, C.M. Will, & A.G. Wiseman, 2004, *Phys. Rev. Lett.* **93**, 091101.

Boercker, B. B., F. J. Rogers, & H. E. DeWitt, 1982, *Phys. Rev. A* **25**, 1623.

Bohm, D. & D. Pines, 1951, *Phys. Rev.* **82**, 625.

Bohm, D. & D. Pines, 1953, *Phys. Rev.* **92**, 609.

Bollinger, J. J., S. L. Gilbert, D. J. Heinzen, W. M. Itano, & D. J. Wineland, 1990, in *Strongly Coupled Plasma Physics*, edited by S. Ichimaru (North-Holland/Yamada Science Foundation, Amsterdam, the Netherlands), p. 117.

Bolton, C. T., 1975, *Astrophys. J.* **200**, 269.

Bowles, K. L., 1958, *Phys. Rev. Lett.* **14**, 45.

Bowles, K. L., 1961, *J. Res. N. B. S.* **65D**, 1.
Bradt, H. et al., 1969, *Nature* **222**, 728.
Brandt, W., A. Ratkowski, & R. H. Ritchie, 1974, *Phys. Rev. Lett* **33**, 1325.
Brittin, W. E., & W. R. Chappell, 1962, *Rev. Mod. Phys* **34**, 620.
Brovman, E. G., Y. Kagan, & A. Kholas, 1972, *Sov. Phys. JETP* **35**, 783.
Brush, S. G., H. L. Sahlin, & E. Teller, 1966, *J. Chem. Phys.* **45**, 2102.
Buneman, O., 1959, *Phys. Rev.* **115**, 503.
Burrows, D. N. et al., 2011, *Nature* **476**, 421.
Callaway, J. & N. H. March, 1984, in *Solid State Physics*, edited by H. Ehrenreich, F. Seitz, & D. Turnbull (Academic Press, New York, NY), Vol. 38, p. 135.
Callen, H. B. & T. A. Welton, 1951, *Phys. Rev.* **83**, 34.
Cameron, A. G. W., 1959, *Astrophys. J.* **130**, 916.
Campanelli, M. C. O. Lousto, P. Marronetti, & Y. Zlochower, 2006, *Phys. Rev. Lett.* **96**, 111101.
Canal, R. & E. Schatzman, 1976, *Astron. Astrophys.* **46**, 229.
Canal, R., J. Isern, & J. Labay, 1990, *Ann. Rev. Astron. Astrophys.* **28**, 183.
Canizares, C. et al., 2005, *Pub. Astron. Soc. Pacif.* **117**, 1144.
Car, R. & M. Parrinello, 1985, *Phys. Rev. Lett.* **55**, 2471.
Castelvecchi, D., 2016, *Nature* **530**, 261.
Ceperley, D. M. & B. J. Alder, 1980, *Phys. Rev. Lett.* **45**, 566.
Ceperley, D. M. & B. J. Alder, 1987, *Phys. Rev. B* **36**, 2092.
Ceperley, D. M. & M. H. Kalos, 1979, in *Monte Carlo Methods in Statistical Physics*, edited by K. Binder (Springer, Berlin, Germany), p. 145.
Chacham, H. & S. G. Louie, 1991, *Phys. Rev. Lett.* **66**, 64.
Chandrasekhar, S., 1935, *Monthly Not. Roy. Astron. Soc.* **95**, 207.
Chandrasekhar, S., 1984, *Revs. Mod. Phys.* **56**, 137.
Chanmugam, G., 1992, *Annu. Rev. Astron. Astrophys.* **30**, 173.
Chatterji, S., L. Blackburn, G. Martin, E. Katsavoundis, 2004, *Class. Quant. Grav.* **21**, S1809.
Cho, B. J. et al., 2011, *Phys. Rev. Lett.* **106**, 167601.
Cicconti, G., D. Frenkel, & I. R. McDonald, 1987, *Simulation of Liquids and Solids* (North-Holland, Amsterdam, the Netherlands).
Clayton, D. D., 1968, *Principles of Stellar Evolution and Nucleosynthesis* (McGraw-Hill, New York, NY).
Cocke, W. J., M. J. Disney, & D. J. Taylor, 1969, *Nature* **221**, 525.
Comella, J. M., H.D Craft, R.V.E. Lovelace, J.M. Sutton , 1969, *Nature* **221**, 1969.
Connaughton, V., 2017, *GCN Circ.* 21506.
D'Antona, F. & I. Mazzitelli, 1990, *Ann. Rev. Astron. Astrophys.* **28**, 139.
Dargarno, A., 1967, *Adv. Chem. Phys.* **12**, 143.
Da Silva, L. B. et al., 1997, *Phys. Rev. Lett.* **78**, 483.
Davidson, R. C., 1990, *Physics of Nonneutral Plasmas* (Addison-Wesley, Redwood City, CA).
Davidson, K. & J. P. Ostriker, 1973, *Astrophys. J.* **179**, 585.
Davies, J. G., A. G. Lyne, & J. H. Seiradakis, 1972, *Nature* **240**, 229.
Debye, P. & E. Hückel, 1923, *Physik. Z.* **24**, 185.
Dhawan, V., I. F. Mirabel, & L. F. Rodriguez, 2000, *Astrophys. J.* **543**, 373.
Dick, R. D. & G. I. Kerley, 1980, *J. Chem. Phys.* **73**, 5264.
Dorchies, D. & V. Recoules, 2016, *Phys. Rep.* **657**, 1.
Dorchies, F. et al., 2008, *Appl. Phys. Lett.* **93**, 121113.

Driscoll, C. F. & J. F. Malmberg, 1983, *Phys. Rev. Lett.* **50**, 167.
Dubin, D. H. E., 1990, *Phys. Rev. A* **42**, 4972.
DuBois, D. F., 1959, *Ann. Phys. (N. Y.)* **8**, 24.
Eardley, D. M. & A. P. Lightman, 1975, *Astrophys. J.* **200**, 187.
Eardley, D. M., A. P. Lightman, & S. L. Shapiro, 1975, *Astrophys. J. Lett.* **199**, L153.
Ebeling, W., A. Förster, V. E. Fortov, V. K. Gryaznov, & A. Ya. Polishchuk, 1991, *Thermodynamic Properties of Hot Dense Plasmas* (Teubner, Stuttgart, Germany).
Eikenberry, S. S., K. Matthews, E.H. Morgan, R.A. Remillard, R.W. Nelson, 1998, *Astrophys. J. Lett.* **494**, L61.
Einstein, A., 1916, *Sitzungsber. K. Preuss. Akad. Wiss.* **1**, 688.
Einstein, A., 1918, *Sitzungsber. K. Preuss. Akad. Wiss.* **1**, 154.
Elphic, R. C. & C. T. Russel, 1978, *Geophys. Res. Lett.* **5**, 211.
Evans, C. R. & C. S. Kochanek, 1989, *Astrophys. J. Lett.* **346**, L13.
Finkelstein, D., 1958, *Phys. Rev.* **110**, 965.
Foldy, L. L., 1978, *Phys. Rev. B* **17**, 4889.
Fortney, J., 2018, *Nature* **555**, 168.
Fortov, V. E., 1982, *Usp. Phys. Nauk* **138**, 361 [*Sov. Phys. Usp.* **25**, 781 (1983)].
Fortov, V. E., 1995, in *Elementary Processes in Dense Plasmas: Proc. Oji International Seminar*, edited by S. Ichimaru & S. Ogata (Addison-Wesley, Reading, MA), p. 389.
Fowler, W. A., 1984, *Rev. Mod. Phys.* **56**, 149.
Fried, B. D. & S. D. Conte, 1961, *The Plasma Dispersion Function* (Academic Press, New York, NY).
Friedel, C. & N. W. Ashcroft, 1977, *Phys. Rev. B* **16**, 662.
Fuchs, K., 1936, *Proc. R. Soc. London A* **153**, 622.
Gamow, G. & E. Teller, 1938, *Phys. Rev.* **53**, 608.
Gaudin, J. et al., 2014, *Sci. Rep.* **4**, 4724.
Goldstein, W., C. Hooper, J. Gauthier, J. Seeley, & R. Lee, eds., 1991, *Radiative Properties of Hot Dense Matter* (World Scientific, Singapore).
Glenzer, S. H. & R. Redmer, 2009, *Rev. Mod. Phys.* **81**, 1625.
Glenzer, S. H. et al., 2007, *Phys. Rev. Lett.* **98**, 065002.
Giacconi, R., H. Gursky, E. M. Kellog, E. Schreier, & H. Tananbaum, 1971, *Astrophys. J. Lett.* **167**, L67.
Gibbons, P. C. S.E. Shatterly, J.J. Risko, & J.R. Fields, 1976, *Phys. Rev. B* **13**, 2451.
Goldreich, P. & W. H. Julian, 1969, *Astrophys. J.* **157**, 869.
Goss, W. M. & U. J. Schwarz, 1971, *Nature Phys. Sci.* **234**, 52.
Graboske, H. C. et al., 1975, *Astrophys. J.* **199**, 265.
Greiner, J., J. G. Cuby, & M. J. McCaughrean, 2001, *Nature* **414**, 522.
Grimes, C. C., 1978, *Surf. Sci.* **73**, 397.
Gudmundsson, E. H., C. J. Pethick, & I. Epstein, 1982, *Astrophys. J. Lett.* **259**, L19.
Guillot, T. et al., 2018, *Nature* **555**, 227.
Hannikainen, D. et al., 2005, *Astron. Astrophys.* **435**, 995.
Hansen, J.-P., 1970, *Phys. Rev. A* **2**, 221.
Hansen, J.-P. & I. R. McDonald, 1986, *Theory of Simple Liquids*, 2nd ed. (Academic Press, London, UK).
Hansen, J.-P. & Verlet, L. 1969, *Phys. Rev.* **184**, 151
Hemley, R. J. et al., 1996, *Phys. Rev. Lett.* **76**, 1667.

Hewish, A., S. J. Bell, J. D. H. Pilkington, P. F. Scott, & R. A. Collins, 1968, *Nature* **217**, 709.
Hirschfelder, J. O., C. F. Curtis, & R. B. Bird, 1954, *Molecular Theory of Gases and Liquids* (Wiley, New York, NY).
Hohenberg, P. & W. Kohn, 1964, *Phys. Rev.* **136**, B864.
Holmes, N. C., M. Ross, & D. A. Young, 1995, *Phys. Rev. B* **52**, 15853.
Holt, S. S., E. A. Boldt, L. J. Kaluzienskii, & P. J. Serlemitsos, 1975, *Nature* **256**, 108.
Hoover, W. G. et al., 1980, *Phys. Rev. A* **22**, 1690.
Hora, H., 1991, *Physics at High Temperature and Density* (Springer, Berlin, Germany).
Hubbard, W. B., 1968, *Astrophys. J.* **152**, 745.
Hubbard, W. B., 1980, *Rev. Geophys. Space Sci.* **18**, 1.
Hubbard, W. B., 1984, *Planetary Interiors* (Van Nostrand Reinhold, New York, NY).
Hubbard, W. B. & M. Lampe, 1969, *Astrophys. J. Suppl.* **18**, 297.
Hubbard, W. B. & M. S. Marley, 1989, *Icarus* **78**, 102.
Hulse, R. A. & J. H. Taylor, 1974, *Astrophys. J.* **191**, L59.
Hulse, R. A. & J. H. Taylor, 1975, *Astrophys. J.* **195**, L51.
Ichimaru, S., 1962, *Ann. Phys.* **20**, 78.
Ichimaru, S., 1970, *Nature* **226**, 731.
Ichimaru, S., 1973, *Basic Principles of Plasma Physics* (W.A. Benjamin, Reading, MA).
Ichimaru, S., 1976, *Astrophys. J.* **208**, 701.
Ichimaru, S., 1977, *Astrophys. J.* **214**, 840.
Ichimaru, S., 1982, *Revs. Mod. Phys.* **54**, 1017.
Ichimaru, S., 1991, *J. Phys. Soc. Jpn.* **60**, 1437.
Ichimaru, S., 1993, *Revs. Mod. Phys.* **65**, 255.
Ichimaru, S., 1996, *Phys. Plasmas* **3**, 233.
Ichimaru, S., 1997, *Phys. Lett. A* **235**, 83.
Ichimaru, S., 2000, *Phys. Rev. Lett.* **84**, 1842.
Ichimaru, S., 2001, *Phys. Plasmas* **8**, 48.
Ichimaru, S., 2004a, *Statistical Plasma Physics—Vol. I: Basic Principles* (Westview Press, Boulder, CO).
Ichimaru, S., 2004b, *Statistical Plasma Physics—Vol. II: Condensed Plasmas* (Westview Press, Boulder, CO).
Ichimaru, S. & H. Kitamura, 1995, *J. Phys. Soc. Jpn.* **64**, 2270.
Ichimaru, S. & H. Kitamura, 1998, in *Proc. Internat'l Conf. Superstrong Fields in Plasmas*, edited by M. Lomtano et al. (AIP Conf. Proc. 426, Varrena, Italy), p. 541.
Ichimaru, S. & H. Kitamura, 1999, *Phys. Plasmas* **6**, 2649; *Erratum* **7**, 3482 (2000).
Ichimaru, S. & S. Ogata, eds., 1995, *Elementary Processes in Dense Plasmas: Proc. Oji International Seminar* (Addison-Wesley, Reading, MA).
Ichimaru, S. & K. Utsumi, 1981, *Phys. Rev. B* **24**, 7385.
Ichimaru, S. & K. Utsumi, 1983, *Astrophys. J. Lett.* **269**, L51.
Ichimaru, S., D. Pines & N. Rostoker, 1962, *Phys. Rev. Lett.* **8**, 231.
Ichimaru, S., H. Iyetomi, & S. Tanaka, 1987, *Phys. Rep.* **149**, 92.
Iess, L. et al., 2018, *Nature* **555**, 220.
Iyetomi, H., S. Ogata, & S. Ichimaru, 1992, *Phys. Rev. A* **46**, 1051.
James, F., 1980, *Rep. Prog. Phys.* **43**, 73.
Jancovici, B., 1962, *Nuovo Cimento* **25**, 428.
Jastrow, R., 1955, *Phys. Rev.* **98**, 1479.

Kadanoff, L. P. & G. Baym, 1962, *Quantum Statistical Mechanics* (W.A. Benjamin, New York, NY).
Kasen, D., B. Metzger, J. Barnes, E. Quataert, E. Ramirez-Ruiz, 2017, *Nature* **551**, 80.
Kaspi, Y. et al., 2018, *Nature* **555**, 223.
Katayama, T. et al., 2013, *Appl. Phys. Lett.* **103**, 135006.
Kennel, C. F. & F. V. Coroniti, 1977, *Ann. Rev. Astron. Astrophys.* **15**, 389.
Kitamura, H. & S. Ichimaru, 1995, *Phys. Rev. E* **51**, 6004.
Kitamura, H. & S. Ichimaru, 1996, *J. Phys. Soc. Jpn.* **65**, 1250.
Kitamura, H. & S. Ichimaru, 1998, *J. Phys. Soc. Jpn.* **67**, 950.
Kivelson, M. G. & D. F. DuBois, 1964, *M. G.* **7**, 1578.
Klein-Wolt, M. et al., 2002, *Mon. Not. R. Astron. Soc.* **331**, 745.
Kohn, W. & L. J. Sham, 1965, *Phys. Rev. A* **140**, 1133.
Kohn, W. & P. Vashishta, 1983, in *Theory of the Inhomogeneous Electron Gas*, edited by S. Lundqvist & N. H. March (Plenum, New York, NY), p. 79.
Krolik, J., 1991, *Astrophys. J. Lett.* **373**, L69.
Kruskal, M. D., 1960, *Phys. Rev.* **119**, 1743.
Kubo, R., 1957, *J. Phys. Soc. Jpn.* **12**, 570.
Lamb, F. K., C. J. Pethick, & D. Pines, 1973, *Astrophys. J.* **184**, 271.
Large, M. I. & A. E. Vaughan, 1972, *Nat. Phys. Sci.* **236**, 117.
Large, M. I., A. E. Vaughan, & B. Y. Mills, 1968, *Nature* **220**, 340.
Landau, L. D., 1956, *Sov. Phys. JETP* **3**, 920.
Landau, L. D., 1957, *Sov. Phys. JETP* **5**, 101.
Landau, L. D. & E. M. Lifshitz, 1960a, *Electrodynamics of Continuous Media* (Addison-Wesley, Reading, MA).
Landau, L. D. & E. M. Lifshitz, 1960b, *Mechanics* (Addison-Wesley, Reading, MA).
Landau, L. D. & E. M. Lifshitz, 1965, *Quantum Mechanics*, 2nd ed. (Addison-Wesley, Reading, MA).
Landau, L. D. & E. M. Lifshitz, 1969, *Statistical Physics*, 2nd ed. (Addison-Wesley, Reading, MA).
Lebowitz, J. L. & J. S. Rowlinson, 1964, *J. Chem. Phys.* **41**, 133.
Levan, A. J. et al., 2011, *Science* **333**, 199.
Lindhard, J., 1954, *Kgl. Danske Videnskab, Selskab Mat.-Fys. Medd.* **28**, 8.
Low, F. J., 1966, *Astron. J.* **71**, 391.
Magro, W. R., D. M. Ceperley, C. Pierleoni, & B. Bernu, 1996, *Phys. Rev. Lett.* **76**, 1240.
Makishima, K., 1995, in *Elementary Processes in Dense Plasmas: Proc. Oji International Seminar*, edited by S. Ichimaru & S. Ogata (Addison-Wesley, Reading, MA), p. 47.
Makishima, K. et al., 1990, *Astrophys. J. Lett.* **365**, L59.
Manchester, R. N. & J. H. Taylor, 1977, *Pulsars* (Freeman, San Francisco, CA).
Manchester, R. N., J. H. Taylor, & G. R. Huguenin, 1975, *Astrophys. J.* **196**, 83.
Manchester, R. N., G.B. Hobbs, A. Teoh, M. Hobbs, 2005, *Astron. J.* **129**, 1993.
Mao, H. K. & R. J. Hemley, 1989, *Science* **244**, 1462.
Mao, H. K. & R. J. Hemley, 1994, *Rev. Mod. Phys.* **66**, 671.
Mao, H. K., R. J. Hemley, & M. Hanfland, 1991, *Phys. Rev. Lett.* **65**, 484.
McDermott, P. N., C. J. Hansen, H. M. Van Horn, & R. Buland, 1985, *Astrophys. J. Lett.* **297**, L37.
McDermott, P. N., H. M. Van Horn, & C. J. Hansen, 1988, *Astrophys. J.* **325**, 725.
McMahon, J. M., M. A. Morales, C. Pierleoni & D. M. Ceperley, 2012, *Revs. Mod. Phys.* **84**, 1607.
McMillan, W. L., 1965, *Phys. Rev. A* **138**, 442.

Menzel, D. H., W. W. Coblentz, & C. O. Lampland, 1926, *Astrophys. J.* **63**, 177.
Mermin, N. D., 1965, *Phys. Rev.* **137**, 1441A.
Metropolis, N., A. W. Rosenbluth, M. N. Rosenbluth, A. H. Teller, & E. Teller, 1953, *J. Chem. Phys.* **21**, 1087.
Michel, F. C., 1982, *Rev. Mod. Phys.* **54**, 1.
Miller, M. C., 2017, *Nature* **551**, 36.
Mirabel, I. F., 1998, *Astron. Astrophys.* **330**, L9.
Mirabel, I. F. & L. F. Rodriguez, 1994, *Nature* **371**, 46.
Mitchell, A. C. & W. J. Nellis, 1981, *Rev. Sci. Instrum.* **52**, 347.
Mostovych, A. N., K. J. Kearney, J. A. Stamper, & A. J. Schmitt, 1991, *Phys. Rev. Lett.* **66**, 612.
Motz, H., 1979, *The Physics of Laser Fusion* (Academic Press, London, UK).
Mroué, A. H. et al., 2013, *Phys. Rev. Lett.* **111**, 241104.
Neilsen, J., & J. C. Lee, 2009, *Nature* **458**, 481.
Nomoto, K., 1982, *Astrophys. J.* **253**, 798 , 780.
Nomoto, K. & Y. Kondo, 1991, *Astrophys. J. Lett.* **367**, L19.
Nomoto, K. & S. Tsuruta, 1981, *Astrophys. J. Lett.* **250**, L19.
Nosé, S. & M. L. Klein, 1983, *Mol. Phys.* **50**, 1055.
Novikov, I. D. & K. S. Thorne, 1973, in *Black Holes,* edited by C. DeWitt & B. DeWitt (Gordon and Breach, New York, NY), p. 343.
Nozières, P. & D. Pines, 1958, *Phys. Rev.* **111**, 442.
Oda, M., P. Gorenstein, H. Gursky, E. Kellogg, E. Schreier, H. Tananbaum, & R. Giacconi, 1971, *Astrophys. J. Lett.* **166**, L1.
Ogata, S. & S. Ichimaru, 1987, *Phys. Rev. A* **36**, 5451.
Ogata, S. & S. Ichimaru, 1989, *Phys. Rev. Lett.* **62**, 2293.
Ogata, S. & S. Ichimaru, 1990, *Phys. Rev. A* **42**, 4867.
Ogata, S., H. Iyetomi, & S. Ichimaru, 1991, *Astrophys. J.* **372**, 259.
Ogata, S., H. Iyetomi, S. Ichimaru, & H. M. Van Horn, 1993, *Phys. Rev. E* **48**, 1344.
O'Neil, T. & N. Rostoker, 1963, *Phys. Fluids* **8**, 1109.
Ortiz, G., M. Harris, & P. Ballone, 1999, *Phys. Rev. Lett.* **82**, 5317.
Ostriker, J. P., & J. E. Gunn, 1969, *Astrophys. J.* **157**, 1394.
Pacini, F., 1967, *Nature* **216**, 567.
Parker, E. N., 1979, *Cosmical Magnetic Fields* (Clarrendon, Oxford, UK).
Pathak, K. N. & P. Vashishta, 1973, *Phys. Rev. B* **7**, 3649.
Pathria, R. K., 1972, *Statistical Mechanics* (Pergamon, Oxford, UK).
Pereira, N. R., J. Davis, & N. Rostoker, eds., 1989, *Dense Z-Pinches* (AIP Conference Proceedings 195, New York, NY).
Perrot, F. & M. W. C. Dharma-wardana, 1984, *Phys. Rev. A* **30**, 2619.
Pian, E. et al., 2017, *Nature* **551**, 67.
Pineo, V. C., L. G. Kraft & H. W. Briscoe, 1960, *J. Geophys. Res.* **65**, 2629.
Pines, D., 1963, *Elementary Excitation in Solids* (W.A. Benjamins New York, NY).
Pines, D. & D. Bohm, 1952, *Phys. Rev.* **85**, 338.
Pines, D. & P. Nozières, 1966, *The Theory of Quantum Liquid* (W. A. Benjamin, New York, NY), Vol. I.
Pollock, E. L. & J.-P. Hansen, 1973, *Phys. Rev. A* **8**, 3110.
Press, W. & K. Thorne, 1972, *Annu. Rev. Astron. Astrophys.* **10**, 335.
Pringle, J. E. & M. J. Rees, 1972, *Astron. Astrophys.* **21**, 1.

Proga, D., 2009, *Nature* **458**, 415.
Raether, H., 1980, *Excitations of Plasmons and Interband Transitions by Electrons* (Springer, Berlin, Germany).
Radhakrishnan, V. & R. N. Manchester, 1969, *Nature* **222**, 228.
Rappaport, S., R. Doxsey, & W. Zauman, 1971, *Astrophys. J. Lett.* **186**, L43.
Ravasio, A. et al., 2007, *Phys. Rev. Lett.* **99**, 135006.
Rees, M. J., 1988, *Nature* **333**, 523.
Reich, E. S., 2013, *Nature* **497**, 296.
Reichley, P. E. & G. S. Downs, 1969, *Nature* **222**, 229.
Rickett, B. J., T. H. Hankins, & J. M. Cordes, 1975, *Astrophys. J.* **201**, 425.
Roberts, D. E. & P. A. Sturrock, 1973, *Astrophys. J.* **181**, 161.
Rosenbluth, M. N. & N. Rostoker, 1962, *Phys. Fluids* **5**, 776.
Ross, M., F. H. Ree, & D. A. Young, 1983, *J. Chem. Phys.* **79**, 1487.
Rostoker, N. & M. N. Rosenbluth, 1960, *Phys. Fluids* **3**, 1.
Rothschild, R. E. et al., 1974, *Astrophys. J. Lett.* **189**, L13.
Ruderman, M. A. & P. G. Sutherland, 1975, *Astrophys. J.* **196**, 51.
Ruoff, A. L. & C. A. Vanderbough, 1990, *Phys. Rev. Lett.* **66**, 754.
Salpeter, E. E., 1952, *Phys. Rev.* **88**, 547.
Salpeter, E. E., 1954, *Aust. J. Phys.* **7**, 373.
Salpeter, E. E. & H. M. Van Horn, 1969, *Astrophys. J.* **155**, 183.
Sanford, P. W., J. C. Ives, S. J. Bell Burnell, K. O. Mason, & P. Murdin, 1975, *Nature* **256**, 109.
Saumon, D., G. Chabrier, & H. M. Van Horn, 1995, *Astrophys. J. Suppl.* **99**, 713.
Sawada, H. et al., 2007, *Phys. Plasmas* **14**, 121702.
Schmidt, G. D. & J. E. Northworthy, 1991, *Astrophys. J.* **366**, 270.
Schreier, E., H. Gursky, E. Kellog, H. Tananbaum, & R. Giacconi, 1971, *Astrophys. J. Lett.* **170**, L21.
Schwarzschild, K., 1916, *Sitzungsber. K. Preuss. Akad. Wiss.* **1**, 189.
Shakura, N. I. & R. A. Sunyaev, 1973, *Astron. Astrophys.* **24**, 337.
Shapiro, S. L. & S. A. Teukolsky, 1983, *Black Holes, White Dwarfs, and Neutron Stars* (Wiley, New York, NY).
Shapiro, S. L., D. M. Eardley, & A. P. Lightman, 1976, *Astrophys. J. Lett.* **204**, 187.
Singwi, K. S., M. P. Tosi, R. H. Land, & A. Sjölander, 1968, *Phys. Rev.* **176**, 589.
Slattery, W. L., G. D. Doolen, & H. E. DeWitt, 1980, *Phys. Rev. A* **21**, 2087.
Slattery, W. L., G. D. Doolen, & H. E. DeWitt, 1982, *Phys. Rev. A* **26**, 2255.
Smartt, S. J. et al., 2017, *Nature* **551**, 75.
Smith, E. J., L. Davis, Jr., & D. E. Jones, 1976, in *Jupiter*, edited by T. Gehrels (Univ. of Arizona Press, Tucson, AZ), p. 788.
Smoluchowski, R., 1967, *Nature* **215**, 691.
Staelin, D. H. & E. C. Reifenstein III, 1968, *Science* **162**, 1481.
Starrfield, S., J. W. Truran, W. M. Sparks, & G. S. Kutter, 1972, *Astrophys. J.* **176**, 169.
Stenlund, E. et al., eds., 1994, *Quark Matter 1993* (North-Holland, Amsterdam, The Netherlands).
Stevenson, D. J., 1980, *J. Phys. (Paris)* **41**, C2–C61.
Stevenson, D. J., 1982, *Ann. Rev. Earth Planet. Sci.* **10**, 257.
Stevenson, D. J. & E. E. Salpeter, 1976, in *Jupiter*, edited by T. Gehrels (Univ. of Arizona Press, Tucson, AZ), p. 85.
Strohmayer, T., S. Ogata, H. Iyetomi, S. Ichimaru, & H. M. Van Horn, 1991, *Astrophys. J.* **375**, 679.

Sturrock, P. A., 1971, *Astrophys. J.* **164**, 529.
Tanaka, S. & S. Ichimaru, 1986, *J. Phys. Soc. Jpn.* **55**, 2278.
Tanaka, S., X. -Z. Yan, & S. Ichimaru, 1990, *Phys. Rev. A* **41**, 5616.
Tananbaum, H., H. Gursky, E. M. Kellog, R. Levinson, R. E. Schreier, & R. Giacconi, 1972a, *Astrophys. J. Lett.* **174**, L143.
Tananbaum, H., H. Gursky, E. M. Kellog, R. Giacconi, & C. Jones 1972b, *Astrophys. J. Lett.* **177**, L5.
Taylor, J. H., R. N. Manchester, & G. R. Huguenin, 1975, *Astrophys. J.* **195**, 513.
Taylor, J. H. & J. M. Weisberg, 1982, *Astrophys. J.* **253**, 908.
The STAR Collaboration, 2017, *Nature* **548**, 62.
Thompson, W. B., 1957, *Proc. Phys. Soc. (London)* **B70**, 1.
Thorne, K. S. & R. H. Price, 1975, *Astrophys. J. Lett.* **195**, L101.
Toigo, F. & T. O. Woodruff, 1971, *Phys. Rev. B* **4**, 371.
Totsuji, H. & S. Ichimaru, 1973, *Prog. Theor. Phys.* **50**, 753.
Troja, E. et al., 2017, *Nature* **551**, 71.
Trümper, J., W. Pietsch, C. Reppin, W. Voges, R. Staubert, E. Kendziorra, 1978, *Astrophys. J. Lett.* **219**, L105.
Tsuneta, S., 1995, in *Elementary Processes in Dense Plasmas: Proc. Oji International Seminar*, edited by S. Ichimaru & S. Ogata (Addison-Wesley, Reading, MA), p. 447.
Utsumi, K. & S. Ichimaru, 1981, *Phys. Rev. B* **23**, 3291.
Vager, Z. & D. S. Gemmel, 1976, *Phys. Rev. Lett.* **37**, 1352.
Van Horn, H. M., 1991, *Science* **252**, 384.
Verschuur, G. L., 1973, *Astrophys. J.* **183**, L9.
Victor, G. A. & A. Dargarno, 1970, *J. Chem. Phys.* **53**, 1316.
Vlasov, A. A., 1967, *Usp. Fiz. Nauk* **93**, 444 [*Soviet Phys. Usp.* **10**, 721].
Voges, W. et al., 1982, *Astrophys. J.* **263**, 803.
Weinberg, S., 1972, *Gravitation and Cosmology* (Wiley, New York, NY).
Weir, S. T., A. C. Mitchell, & W. J. Nellis, 1996, *Phys. Rev. Lett.* **76**, 1860.
Weisheit, J., 1995, in *Elementary Processes in Dense Plasmas: Proc. Oji International Seminar*, edited by S. Ichimaru & S. Ogata (Addison-Wesley, Reading, MA), p. 61.
Whelan, J. & I. Iben, 1973, *Astrophys. J.* **186**, 1007.
Wigner, E. P., 1932, *Phys. Rev.* **40**, 749.
Wigner, E. P., 1935, *J. Chem. Phys.* **3**, 764.
Wigner, E. P., 1938, *Trans. Faraday Soc.* **34**, 678.
Wigner, E. & H. B. Huntington, 1935, *J. Chem. Phys.* **3**, 764.
Yagi, K., T. Hatsuda & Y. Miake, 2005, *Quark Gluon Plasma* (Cambridge Univ. Press, Cambridge, UK).
Yonezawa, F., ed., 1992, *Molecular Dynamics Simulations* (Springer, Berlin, Germany).
Zauderer, B. A. et al., 2011, *Nature* **476**, 425.

INDEX

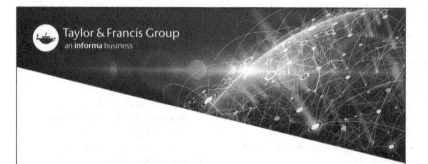